Analysis of
Classic Programming Cases in Android

Android 编程
经典案例解析

高成珍 钟元生 高必梵 何英 编著

清华大学出版社
北京

内 容 简 介

本书为Android编程初学者提高、自测和加入开发团队提供贴心帮助，内容包括常用案例、常见上机调试错误、Android程序员猎头系统和自测题。

本书所涉及的Android经典案例效果，读者稍加改动就可直接应用于自己的项目中，包括TextView特效、手机屏幕区域划分、我的课表（表格布局应用）、闪烁霓虹灯（层布局应用）、简易计算器设计（布局综合运用）、页面滑动切换效果、图片定时滑动播放效果、搜索关键字提示、仿画廊视图效果、城市景点介绍、高校新闻（延迟加载效果、下拉刷新效果、选项卡切换效果）、省市二级列表（ExpandableListView应用）、产品分类（自定义多级列表效果）、天气预报（Web Service调用）和音乐之声（音乐播放器）等。

为引导读者理解、掌握和灵活运用每个案例，编者通过图解分析、代码展示、技术剖析，由浅入深引导读者将所学知识融会贯通。本书既可作为Android入门者的提高练习指南，又可作为移动开发者的好帮手。

本书封面贴有清华大学出版社防伪标签，无标签者不得销售。
版权所有，侵权必究。侵权举报电话：010-62782989　13701121933

图书在版编目（CIP）数据

Android编程经典案例解析/高成珍，钟元生等编著．—北京：清华大学出版社，2015
ISBN 978-7-302-38293-5

Ⅰ．①A…　Ⅱ．①高…　②钟…　Ⅲ．①移动终端－应用程序－程序设计　Ⅳ．①TN929.53

中国版本图书馆CIP数据核字（2014）第241353号

责任编辑：袁勤勇　王冰飞
封面设计：傅瑞学
责任校对：白　蕾
责任印制：沈　露

出版发行：清华大学出版社
　　　　　网　　址：http://www.tup.com.cn, http://www.wqbook.com
　　　　　地　　址：北京清华大学学研大厦A座　　　　邮　编：100084
　　　　　社 总 机：010-62770175　　　　　　　　　　邮　购：010-62786544
　　　　　投稿与读者服务：010-62776969, c-service@tup.tsinghua.edu.cn
　　　　　质量反馈：010-62772015, zhiliang@tup.tsinghua.edu.cn
　　　　　课件下载：http://www.tup.com.cn, 010-62795954
印 装 者：三河市少明印务有限公司
经　　销：全国新华书店
开　　本：185mm×260mm　　印　张：19.25　　字　数：447千字
版　　次：2015年1月第1版　　　　　　　　　印　次：2015年1月第1次印刷
印　　数：1～2000
定　　价：39.50元

产品编号：061810-01

前言

随着 Android 手机的普及、Android 应用的生活化,特别是社会上 Android 人才需求旺盛、Android 工程师薪酬丰厚的状况,越来越多的院校开始关注对 Android 人才的培养,开设手机编程课是大多数高校计算机相关专业未来几年的必然选择。

为此,在江西省大学生手机软件设计赛指导教师的 Android 编程培训讲义的基础上,我们编写了《Android 应用开发教程》,该书于 2013 年 1 月在江西高校出版社出版。该书出版后被江西省的多所本、专科院校选用,例如江西师范大学、江西财经大学、东华理工大学、江西科技师范大学、井冈山大学、赣南师范学院、九江学院等本科院校,江西应用技术职业学院、南昌工学院、江西环境工程职业学院等专科院校。另外,江西省外的天津中德职业技术学院、厦门理工学院的软件学院等多所高校也选用了该书。同时,该书被选为培训用书,如南昌大学软件学院的暑期培训、南昌易游培训学校的 Android 培训等选用了该书。该书的出版带动了部分高校开设 Android 相关课程,同时吸引了一批网友的关注。

许多教师和网友反映,该书实用、通俗易懂、深入浅出、可读性强,特别适合大学课堂教学和初学者入门自学。并且,希望我们再出版一本常用案例分析教材,针对 Android 实际开发中经常使用到的功能或效果进行解剖,从而提高大家综合运用知识的能力。

在调研多家企业对 Android 研发的相关岗位的需求时,许多企业纷纷表示希望与我们合作,让我们代为招聘和测试 Android 开发人员。基于此,我们开发了一套 Android 程序员代招代测系统,并制定了一套测试体系,包括初级、中级、高级不同层次。为了让测试者明确测试内容及相关技能,我们还提供了一些典型案例作为参考。

为了提高学生运用 Android 的能力,检验学生是否掌握了相应的基本技能,同时为了能够帮企业招聘到具有一定项目经验、能够立即参与项目开发的 Android 开发人员,我们结合自身的高校教学经验及实际的 Android 项目开发经历,通过细致的整理和分析,对专业技能和基本知识进行了合理划分,最终设计和编写了这本《Android 编程经典案例解析》。本书案例是在原有知识的基础上添加了一些功能和新的效果,主要检验学生是否能够灵活地运用所学内容,以及是否掌握了 Android 的学习方法,是否具备自学能力。在设计这些案例时,主要考虑了以下几个方面。

(1) 实用:模拟 Android 应用开发中经常使用到的功能和效果。

(2) 综合性:每个案例都涉及多方面的知识点,需灵活运用。

(3) 注重案例分析:网络上的 Android 源代码虽然非常多,但详细分析开发过程的较少,再加上注释少、编码风格不同,很多案例下载之后难以为自己所用,本书在编写时注重了对案例的详细分析。

本书详细地分析了 17 个典型 Android 案例的开发过程,这些案例紧贴市场,实用价值高,读者稍加修改便可用于自己的项目当中。本书同时介绍了 Android 开发中常见的错误和程序调试方法,并提供了相应的 Android 测试题。在学完本书之后,读者能够具备独立进行项目开发的能力。

本书由高成珍担任主编,负责全书案例的选取和大部分章节的编写工作;钟元生担任联合主编,具体负责编写指导、体例设计、编撰组织、审稿和质量保证工作。本书各章的分工如下:高成珍负责第5~18章;钟元生负责第1、第2和第19章;高必梵负责第3章;何英负责第4章,并参与第19章的编写。另外,研究生杨旭、章雯、陈海俊、吴微微、黄婧、曹权等参与了本书的初稿讨论以及配套教学课件的制作工作。

本书在编写过程中得到了江西财经大学软件与通信工程学院、江西科技学院信息工程学院、南昌倚动软件有限公司、江西机电职业技术学院校企合作办公室的帮助与支持,在此一并致谢。

由于编者技术水平有限,再加上时间仓促,书中不足之处在所难免,希望广大读者多提宝贵意见。

编　者

2014 年 10 月

阅 读 指 南

本书假定您已接触过一些 Android 基础知识,知道 Android 应用程序的基本结构和一些常见的界面控件。如果尚未接触过 Android 应用开发,请自学我们编写的《Android 应用开发教程》教材,或者参考我们录制的手把手教你学 Android 4.1 系列视频,资源网址为 http://10lab.cn/resource.html。

书中示例较多,源代码较长。本书注重示例的程序分析,为了方便大家掌握知识重点,压缩了相应篇幅,文中仅列出了一些关键代码,读者可到 http://10lab.cn/case/resource.html 下载完整源码,直接运行即可看到书中的效果。

强烈建议,您在阅读本书时,**自己根据书中解释和关键代码补充完成完整程序**,而不是直接打开源程序直接运行观看结果。仅在反复尝试失败时,才参考提供的源码,不建议一开始就看程序源码。

为了方便教师教学,书中为每一段源码分别添加了行号,并为一些关键语句添加了注释,例如:

```
1   public class MainActivity extends Activity {
2       public void onCreate(Bundle savedInstanceState) {
3           super.onCreate(savedInstanceState);      //调用父类的该方法
4           setContentView(R.layout.activity_main);  //设置Activity对应的界
                                                     //面布局文件
5       }
6       public boolean onCreateOptionsMenu(Menu menu) {    //创建选项菜单
7           getMenuInflater().inflate(R.menu.activity_main, menu);//指定菜单资源
8           return true;
9       }
10  }
```

其中,左边的 1、2、3、…表示行号,中间的"super.onCreate(savedInstanceState);"才是真实的程序代码内容,符号"//"及后面的内容表示对中间代码的注释。

为了方便学习、交流、资源共享,我们提供了相应资源的网络下载地址,其中有源码、课件、试题等,网址为 http://10lab.cn/case/resource.html。

如果大家在学习或使用本书过程中有什么疑问或有什么好的建议,欢迎大家通过 QQ 群 314753495 或邮箱 1281147324@qq.com 与我们联系。

目录

第 1 章　TextView 特效　<<< 1
- 1.1　案例概述 …………………………………………………………… 1
- 1.2　关键代码 …………………………………………………………… 2
- 1.3　代码分析 …………………………………………………………… 3
  - 1.3.1　TextView 中文字滚动的效果 ………………………………… 3
  - 1.3.2　同一 TextView 中文字颜色不同的效果 …………………… 4
  - 1.3.3　TextView 中文字周围图片环绕的效果 …………………… 4
  - 1.3.4　自动链接效果 ………………………………………………… 5
- 1.4　知识扩展 …………………………………………………………… 5
  - 1.4.1　android:gravity 与 android:layout_gravity 的区别 ……… 5
  - 1.4.2　android:padding 与 android：layout_margin 的区别 …… 5
  - 1.4.3　Android 中颜色值的表示 …………………………………… 6
- 1.5　思考与练习 ………………………………………………………… 7

第 2 章　手机屏幕的区域划分　<<< 8
- 2.1　案例概述 …………………………………………………………… 8
- 2.2　关键代码 …………………………………………………………… 9
- 2.3　代码分析 …………………………………………………………… 10
  - 2.3.1　线性布局 ……………………………………………………… 10
  - 2.3.2　按比例分割屏幕 ……………………………………………… 10
- 2.4　知识扩展 …………………………………………………………… 11
- 2.5　思考与练习 ………………………………………………………… 11

第 3 章　我的课表——表格布局的应用　<<< 13
- 3.1　案例概述 …………………………………………………………… 13
- 3.2　关键代码 …………………………………………………………… 13
- 3.3　代码分析 …………………………………………………………… 18
  - 3.3.1　界面分析 ……………………………………………………… 18
  - 3.3.2　表格布局 ……………………………………………………… 18
  - 3.3.3　为 TextView 添加边框 ……………………………………… 19
  - 3.3.4　定义样式 ……………………………………………………… 19
  - 3.3.5　直接绑定到标签 ……………………………………………… 20

3.4　知识扩展 ··· 20
　　3.5　思考与练习 ··· 21

第 4 章　闪烁霓虹灯——层布局的应用　　<<< 22
　　4.1　案例概述 ··· 22
　　4.2　关键代码 ··· 22
　　4.3　代码分析 ··· 24
　　　　4.3.1　界面分析 ··· 24
　　　　4.3.2　相对布局 ··· 25
　　　　4.3.3　层布局 ·· 25
　　　　4.3.4　定时器 ·· 26
　　　　4.3.5　Handler 消息传递 ·· 26
　　4.4　知识扩展 ··· 27
　　4.5　思考与练习 ··· 28

第 5 章　简易计算器——布局的综合应用　　<<< 29
　　5.1　案例概述 ··· 29
　　5.2　关键代码 ··· 29
　　5.3　代码分析 ··· 34
　　　　5.3.1　界面分析 ··· 34
　　　　5.3.2　网格布局 ··· 34
　　5.4　知识扩展 ··· 38
　　5.5　思考与练习 ··· 43

第 6 章　页面滑动切换　　<<< 44
　　6.1　案例概述 ··· 44
　　6.2　关键代码 ··· 44
　　6.3　代码分析 ··· 52
　　　　6.3.1　界面分析 ··· 52
　　　　6.3.2　ViewPager 介绍 ·· 52
　　6.4　知识扩展 ··· 53
　　　　6.4.1　基于监听的事件处理 ·· 53
　　　　6.4.2　页面全屏显示 ··· 55
　　6.5　思考与练习 ··· 55

第 7 章　图片定时滑动播放效果　　<<< 56
　　7.1　案例概述 ··· 56
　　7.2　关键代码 ··· 56

7.3 代码分析 61
　　7.3.1 界面分析 61
　　7.3.2 自定义 MyImageTopView 控件 62
7.4 知识扩展 62
　　7.4.1 自定义控件 62
　　7.4.2 手势检测 63
7.5 思考与练习 64

第 8 章　智能提示　<<< 65

8.1 案例概述 65
8.2 关键代码 65
8.3 代码分析 68
　　8.3.1 智能提示完成输入 68
　　8.3.2 智能更新数据源 69
8.4 知识扩展 69
　　8.4.1 ArrayAdapter 介绍 69
　　8.4.2 对话框 70
8.5 思考与练习 71

第 9 章　仿画廊视图效果　<<< 72

9.1 案例概述 72
9.2 关键代码 72
9.3 代码分析 75
　　9.3.1 界面分析 75
　　9.3.2 ImageSwitcher 介绍 75
9.4 知识扩展 76
9.5 思考与练习 77

第 10 章　南昌景点介绍　<<< 78

10.1 案例概述 78
10.2 关键代码 78
10.3 代码分析 83
　　10.3.1 界面分析 83
　　10.3.2 ListView 介绍 84
　　10.3.3 SimpleAdapter 介绍 85
　　10.3.4 ClipDrawable 介绍 86
10.4 知识扩展 87
　　10.4.1 raw 目录介绍 87

10.4.2 Activity 概述 ································· 87
10.5 思考与练习 ····································· 89

第 11 章 财大新闻——ListView 延迟加载效果 <<< 90

11.1 案例概述 ····································· 90
11.2 关键代码 ····································· 90
11.3 代码分析 ····································· 95
  11.3.1 ListView 延迟加载原理 ················ 95
  11.3.2 SQLite 数据库介绍 ··················· 96
11.4 知识扩展 ····································· 100
11.5 思考与练习 ····································· 100

第 12 章 财大新闻——ListView 下拉刷新效果 <<< 102

12.1 案例概述 ····································· 102
12.2 关键代码 ····································· 103
12.3 代码分析 ····································· 111
12.4 知识扩展 ····································· 112
12.5 思考与练习 ····································· 113

第 13 章 学院介绍——选项卡切换效果 <<< 114

13.1 案例概述 ····································· 114
13.2 关键代码 ····································· 114
13.3 代码分析 ····································· 122
  13.3.1 TabHost 介绍 ······················ 122
  13.3.2 Fragment 介绍 ····················· 123
  13.3.3 根据状态改变图片 ················ 124
13.4 知识扩展 ····································· 125
  13.4.1 Fragment 与 Activity 交互 ············ 125
  13.4.2 ActionBar 实现页面切换效果 ········ 129
13.5 思考与练习 ····································· 131

第 14 章 省市二级列表——ExpandableListView 的应用 <<< 132

14.1 案例概述 ····································· 132
14.2 关键代码 ····································· 132
14.3 代码分析 ····································· 135
14.4 知识扩展 ····································· 136
14.5 思考与练习 ····································· 139

第 15 章　产品分类——自定义多级列表效果　<<< 140

- 15.1　案例概述 ························· 140
- 15.2　关键代码 ························· 140
- 15.3　代码分析 ························· 146
- 15.4　知识扩展 ························· 147
- 15.5　思考与练习 ······················· 153

第 16 章　天气预报——Web Service 的调用　<<< 154

- 16.1　案例概述 ························· 154
- 16.2　关键代码 ························· 155
- 16.3　代码分析 ························· 178
  - 16.3.1　调用 Web Service ··············· 178
  - 16.3.2　用 SharedPreference 保存用户信息 ···· 180
  - 16.3.3　按两次返回键退出应用程序 ········· 181
- 16.4　知识扩展 ························· 181
- 16.5　思考与练习 ······················· 182

第 17 章　音乐播放器　<<< 183

- 17.1　案例概述 ························· 183
- 17.2　关键代码 ························· 184
- 17.3　代码分析 ························· 215
  - 17.3.1　音乐播放器的主要功能分析 ········· 215
  - 17.3.2　Android 四大组件之 ContentProvider ···· 216
  - 17.3.3　Android 四大组件之 Service ········ 218
  - 17.3.4　Android 四大组件之 BroadcastReceiver ·· 219
- 17.4　知识扩展 ························· 221
  - 17.4.1　媒体播放器 MediaPlayer ·········· 221
  - 17.4.2　发送通知 Notification ············ 224
- 17.5　思考与练习 ······················· 225

第 18 章　Android 中常见的错误与程序调试方法　<<< 227

- 18.1　程序调试工具 ····················· 227
  - 18.1.1　LogCat 工具介绍 ··············· 227
  - 18.1.2　Eclipse 提供的 Debug 功能 ········ 230
- 18.2　运行时常见的错误 ················· 230
  - 18.2.1　空指针异常 ·················· 230
  - 18.2.2　类型转换异常 ················· 233
  - 18.2.3　数组越界异常 ················· 233

18.2.4 重复运行程序出现警告 ……………………………………………… 233
18.2.5 XML 文件中标签拼写错误 …………………………………………… 234
18.2.6 使用 ListActivity 时调用 setContentView()方法出错 ………… 234
18.2.7 在 Eclipse 中导入项目时错误 ……………………………………… 235

第 19 章 Android 程序员猎头系统 <<< 236

19.1 系统功能概述………………………………………………………………… 236
19.2 系统结构……………………………………………………………………… 236
  19.2.1 开发技术………………………………………………………………… 236
  19.2.2 主页面介绍……………………………………………………………… 237
  19.2.3 系统功能流程图………………………………………………………… 237
19.3 系统业务操作流程…………………………………………………………… 239
  19.3.1 企业招聘操作流程……………………………………………………… 239
  19.3.2 应聘者求职操作流程…………………………………………………… 244
  19.3.3 社交化测试流程………………………………………………………… 253
19.4 系统角色使用流程…………………………………………………………… 258
  19.4.1 企业用户操作流程……………………………………………………… 258
  19.4.2 应聘者操作流程………………………………………………………… 263
  19.4.3 评委操作流程…………………………………………………………… 270
  19.4.4 超级管理员操作流程…………………………………………………… 273

附录 A Android 编程测试题 <<< 280

TextView 特效

1.1 案例概述

在 Android 应用中，TextView 是使用最多的控件之一，主要用于展示文本。为了突出某些主题，往往需要以不同的颜色或加粗等方式显示主题。此外，手机屏幕宽度有限，对于文本内容过长而又不想影响其他内容，较好的解决方案是以滚动方式显示。本案例主要展示 TextView（文本显示框）的一些特殊效果。例如，当文本内容较多时文本滚动显示的效果、同一文本显示框中设置多种文本颜色的效果、文本周围显示图片的效果、自动识别文本中的各种链接效果等。通过本案例的学习，读者可以轻松地实现界面图文并茂、文本滚动等效果。本案例的程序运行效果如图 1-1 所示，界面分析如图 1-2 所示。

图 1-1　程序运行效果图

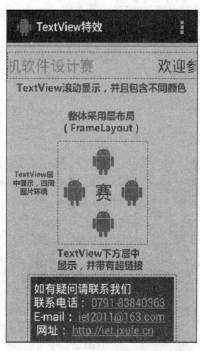

图 1-2　程序运行界面分析图

1.2 关键代码

布局文件：01\TextViewEffect\res\layout\activity_main.xml

```xml
1   <FrameLayout xmlns:android="http://schemas.android.com/apk/res/android"
2       xmlns:tools="http://schemas.android.com/tools"
3       android:layout_width="match_parent"
4       android:layout_height="match_parent"
5       android:background="#aabbcc">
6       <TextView
7           android:id="@+id/title"
8           android:layout_width="wrap_content"
9           android:layout_height="wrap_content"
10          android:ellipsize="marquee"
11          android:focusable="true"
12          android:focusableInTouchMode="true"
13          android:singleLine="true"
14          android:textColor="#0000ff"
15          android:textSize="24sp"
16          android:layout_marginTop="20dp"/>
17      <TextView
18          android:layout_width="wrap_content"
19          android:layout_height="wrap_content"
20          android:layout_gravity="center"
21          android:gravity="center"
22          android:textSize="30sp"
23          android:text="@string/content"
24          android:drawableTop="@drawable/ic_launcher"
25          android:drawableBottom="@drawable/ic_launcher"
26          android:drawableLeft="@drawable/ic_launcher"
27          android:drawableRight="@drawable/ic_launcher"
28          android:textColor="#0000ff"/>
29      <TextView
30          android:layout_width="wrap_content"
31          android:layout_height="wrap_content"
32          android:text="@string/info"
33          android:textColor="#ffffff"
34          android:textSize="18sp"
35          android:autoLink="all"
36          android:layout_gravity="bottom|center_horizontal"
37          android:background="#0000ff"
38          android:padding="5dp"/>
39  </FrameLayout>
```

字符串常量文件：01\TextViewEffect\res\values\strings.xml

```
1  <?xml version="1.0" encoding="utf-8"?>
2  <resources>
3      <string name="app_name">TextView 特效</string>
4      <string name="action_settings">Settings</string>
5      <string name="title">欢迎参加江西省大学生 &lt;font color=red&gt;
            手机软件设计赛 &lt;/font&gt;</string>
6      <string name="content">赛</string>
7      <string name="info">如有疑问请联系我们\n 联系电话：0791-83840363
            \nE-mail: iet2011@163.com
            \n 网址：http://iet.jxufe.cn </string>
8  </resources>
```

程序代码：01\TextViewEffect\src\iet\jxufe\cn\android\textvieweffect\MainActivity.java

```
1  public class MainActivity extends Activity {
2      private TextView mTitle;                              //显示标题的文本框
3      @Override
4      protected void onCreate(Bundle savedInstanceState) {
5          super.onCreate(savedInstanceState);
6          setContentView(R.layout.activity_main);           //设置显示界面
7          mTitle=(TextView)findViewById(R.id.title);        //根据 ID 在布局文件
                                                             //中查找控件
8          mTitle.setText(Html.fromHtml(getResources().getString
              (R.string.title)));                            //为文本显示框设置文本内容
9      }
10     @Override
11     public boolean onCreateOptionsMenu(Menu menu){
12         getMenuInflater().inflate(R.menu.main, menu);
13         return true;
14     }
15 }
```

1.3 代码分析

1.3.1 TextView 中文字滚动的效果

要想实现文字滚动显示效果，TextView 需要满足以下条件。

(1) 文本显示框的内容超过文本显示框的宽度。

(2) 文本显示框设置单行显示，即 **android：singleLine＝"true"**。默认情况下，超出宽度的文字会自动换行显示。

(3) 设置文本滚动显示，即 **android：ellipsize＝"marquee"**。在单行情况下，超出文本

显示框宽度的部分默认不显示。

（4）TextView 需要获取焦点，设置 android：focusableInTouchMode、android：focusable 属性的值为 true，当文本显示框失去焦点时，文本将不再滚动。

1.3.2 同一 TextView 中文字颜色不同的效果

在 Android 中，对控件显示效果的控制有两种方式，一种是在 XML 文件中通过 XML 标签的一些属性来控制控件显示。另一种是在 Java 代码中通过调用控件相应类的方法来控制控件显示。关于 TextView 中文字的颜色，在 XML 文件中可以通过 android：textColor 属性设置文本的颜色，但此时设置的是所有文字的颜色，即同一 TextView 中文字的颜色始终是一致的，通过 android:textColor 属性无法实现同一文本框中显示多种颜色文字的效果，因此可考虑通过 Java 代码来实现该效果。在 Android 中对 HTML 有良好的支持，通过 HTML 类的一些静态方法可以对字符串中的 HTML 标签进行解析，因此只需将 TextView 中的文字设置为包含 HTML 中颜色标签的字符串，然后通过 HTML 类解析即可。

Java 代码和 XML 布局文件是相互独立的两个文件，在 Android 中通过 setContentView 方法将 XML 布局文件和 Java 代码关联起来。在 XML 布局中可能存在多个控件，若想在 Java 代码中控制某个具体的控件，首先需要获取到该控件。在 Android 中是通过为控件添加 ID 属性来唯一标记该控件的，因此需要为 TextView 控件添加 ID 属性，即 android:id="@+id/title"，然后在代码中调用 findViewById(R.id.title)得到该控件，最后调用 TextView 的 setText()方法来设置文本内容。在此调用了 HTML 类的静态方法 fromHtml()对字符串进行解析，主要是将 HTML 标签转化成相应的显示格式。这个字符串是 R.string.title 资源 ID 所指引的资源，即 strings.xml 文件中的"欢迎参加江西省大学生 <font color=red>手机软件设计赛 </font>"，其中，<代表"<"，>代表">"，这是因为在 XML 文件中"<"、">"有特殊含义，在字符串中不能直接包含这些字符。实际上，也可以直接在代码中表示，即：

    mTitle.setText(Html.fromHtml("欢迎参加江西省大学生<font color=red>手机软件设计赛</font>"));

其中，<font color=red></font>是 HTML 中的标签，解析结果就是将<font>标签内的文字颜色设置为红色。

1.3.3 TextView 中文字周围图片环绕的效果

在 Android 中，TextView 上不仅可以显示文字，还可以显示图片。TextView 控件提供了 android:drawableTop、android:drawableBottom、android:drawableLeft、android:drawableRight 等属性用于指定图片的方位，即位于文字的上方、下方、左边、右边等，同时还提供了 android:drawablePadding 属性用于设置图片与文字之间的边距，通过这些属性可以很方便地实现一些简单的图文并茂的效果。对于一些复杂的图文并茂效果，可以通过综合使用 TextView 与 ImageView 来实现。

注意：当图片大小与文字大小不一致时，为了使显示美观，可设置文字的对齐方式。

1.3.4 自动链接效果

在 Android 应用中，特别是一些宣传页面中，经常会用到链接，单击链接后可以调用系统中相应的程序来执行相关操作，例如网址链接、电话链接、邮箱链接等。其实这些链接在 Android 中实现非常简单，只需要设置 TextView 的 android:autoLink 属性即可，该属性有以下几个值。

- none：不匹配任何格式，这是默认值。
- web：只匹配网址，文本中的网址会以超链接的形式显示。
- email：只匹配电子邮箱，文本中的电子邮箱会以超链接的形式显示。
- phone：只匹配电话号码，文本中的电话号码会以超链接的形式显示。
- map：只匹配地图地址，文本中的地址会以超链接的形式显示。
- all：匹配以上所有值。

除了该方法之外，用户也可以通过解析 HTML 标签的方式对网址、E-mail、电话等添加超链接，但此时仅仅有超链接显示效果，单击后并不会调用相关的程序执行该操作。

1.4 知识扩展

1.4.1 android:gravity 与 android:layout_gravity 的区别

在 activity_main.xml 文件的第 2 个 TextView 中（第 20 行和第 21 行）既设置了 android:gravity="center" 又设置了 android:layout_gravity="center"，这两个属性都是用来设置对齐方式，那么它们有什么区别呢？

android:gravity 表示的是控件里面内容的对齐方式，而 android:layout_gravity 表示的是整个控件在它的父容器中的对齐方式。两个属性的效果分析如图 1-3 所示。

图 1-3 两种对齐方式的效果分析图

1.4.2 android:padding 与 android:layout_margin 的区别

在 Android 中，android:padding 与 android:layout_margin 属性都是用来设置边距

大小，它们有什么区别呢？

android:padding 表示的是控件里面的内容距离控件边缘的长度，当控件的大小值为 wrap_content 时，在设置 padding 的值时，控件将会变大，即 padding 属于控件的一部分，而 android:layout_margin 表示的是整个控件与它的父容器或者兄弟控件之间的距离。当为控件设置背景时，背景会填充 padding 部分，但不会填充 margin 部分。padding 表示的是控件内的距离关系，margin 表示的是控件间的距离关系。两种设置边距方式的效果分析如图 1-4 所示。

图 1-4 两种设置边距方式的效果分析图

注意：android:padding 和 android:layout_margin 设置的边距是指上、下、左、右的边距，如果仅需要设置某一方向的边距，例如左边距，可以使用 android:paddingLeft 或者 android:layout_marginLeft。

1.4.3 Android 中颜色值的表示

在 Android 中经常需要使用到颜色，例如设置文本的颜色、控件的背景颜色等。在 XML 文件中颜色的表示有两种方式，一种是通过十六进制数来表示，另一种是通过引用系统中提供的一些颜色常量来表示。在通过十六进制数来表示时，颜色值总是以♯号开头，通过红（Red）、绿（Green）、蓝（Blue）三原色以及一个透明度（Alpha）值来表示，如果省略了透明度的值，那么该颜色默认是完全不透明的，因此，颜色表示主要有以下几种。

- ♯RGB：用 3 位十六进制数表示颜色，R 表示红色，G 表示绿色，B 表示蓝色，每种颜色的值为 0~f，共 16 级。
- ♯RRGGBB：用 6 位十六进制数表示颜色，RR 表示红色，GG 表示绿色，BB 表示蓝色，每种颜色的值为 00~ff，共 256 级。
- ♯ARGB：用 4 位十六进制数表示颜色，A 表示透明度，R 表示红色，G 表示绿色，B 表示蓝色，每种颜色的值为 0~f，共 16 级。
- ♯AARRGGBB：用 8 位十六进制数表示颜色，AA 表示透明度，RR 表示红色，GG 表示绿色，BB 表示蓝色，每种颜色的值为 00~ff，共 256 级。

在 Android 的 XML 文件中，引用 Android 系统中颜色的方式为"@android:color/颜色"，例如"@android:color/holo_red_dark"表示深红色。

除此之外，在 Android 的程序代码中，可通过 Color 类来定义各种颜色。

1.5 思考与练习

（1）本案例使用了层布局(FrameLayout)，用线性布局(LinearLayout)能不能实现同样的效果？如果能，请尝试实现；如果不能，请说明理由。

（2）尝试综合使用 TextView 与 ImageView 实现文字周围图片环绕的效果（提示：使用相对布局）。

（3）在层布局(FrameLayout)中，每个控件会单独占一层，因此在 FrameLayout 标签中不包含 android:gravity 属性，也不包含 android:layout_gravity 属性。请判断这个说法是否正确。

（4）在层布局(FrameLayout)中包含一个按钮(Button)，如果让该按钮在其父容器中居中显示，正确的方法是（　　）。

　　A）设置按钮的属性：android:layout_gravity="center"

　　B）设置按钮的属性：android:gravity="center"

　　C）设置按钮父容器的属性：android:layout_gravity="center"

　　D）设置按钮父容器的属性：android:gravity="center"

（5）在以下选项中，不能合法表示颜色值的是（　　）。

　　A）#ggg　　　　B）#ffff　　　　C）#eeeeee　　　　D）#dddddddd

（6）在以下选项中，不能合法表示颜色值的是（　　）。

　　A）#aaa　　　　B）#aaaa　　　　C）#aaaaa　　　　D）#aaaaaa

第 2 章

手机屏幕的区域划分

2.1 案例概述

在实际应用中，大家经常会遇到分割屏幕的需求，例如两个控件的宽度比为 1∶3，或者某个控件的宽度占总屏幕宽度的 1/3 等。由于不同手机的屏幕大小会有所不同，因此不能直接计算出 1∶3 或 1/3 各是多少，然后在布局中使用固定的像素值。在 Android 中提供了 android:layout_weight 属性，通过该属性可以很容易地实现按比例分配空间的效果。本案例主要介绍 android:layout_weight 属性的用法，实现将屏幕垂直分割成上、中、下 3 个部分的效果，它们的高度比为 1∶4∶1，然后将中间部分又分割为左、中、右 3 个部分，它们的宽度比为 1∶4∶1。本案例的程序运行效果如图 2-1 所示，界面分析如图 2-2 所示。

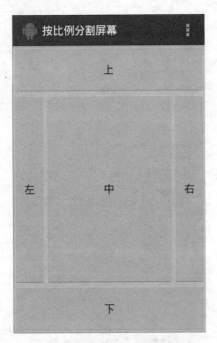

图 2-1 程序运行效果图

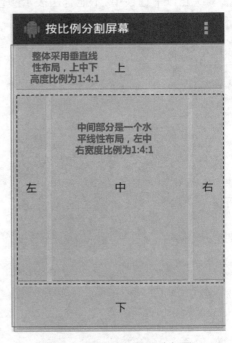

图 2-2 程序运行界面分析图

2.2 关键代码

布局文件：02\DivideScreen\res\layout\activity_main.xml

```xml
 1  <LinearLayout xmlns:android="http://schemas.android.com/apk/res/android"
 2      xmlns:tools="http://schemas.android.com/tools"
 3      android:layout_width="match_parent"
 4      android:layout_height="match_parent"
 5      android:background="#aabbcc"
 6      android:orientation="vertical">
 7      <Button
 8          android:layout_width="match_parent"
 9          android:layout_height="0dp"
10          android:layout_weight="1"
11          android:text="@string/up"/>
12      <LinearLayout
13          android:layout_width="match_parent"
14          android:layout_height="0dp"
15          android:layout_weight="4"
16          android:orientation="horizontal">
17          <Button
18              android:layout_width="0dp"
19              android:layout_height="match_parent"
20              android:layout_weight="1"
21              android:text="@string/left"/>
22          <Button
23              android:layout_width="0dp"
24              android:layout_height="match_parent"
25              android:layout_weight="4"
26              android:text="@string/center"/>
27          <Button
28              android:layout_width="0dp"
29              android:layout_height="match_parent"
30              android:layout_weight="1"
31              android:text="@string/right"/>
32      </LinearLayout>
33      <Button
34          android:layout_width="match_parent"
35          android:layout_height="0dp"
36          android:layout_weight="1"
37          android:text="@string/down"/>
38  </LinearLayout>
```

字符串常量文件：02\DivideScreen\res\values\strings.xml

```
1   <?xml version="1.0" encoding="utf-8"?>
2   <resources>
3       <string name="app_name">按比例分割屏幕</string>
4       <string name="action_settings">Settings</string>
5       <string name="up">上</string>
6       <string name="down">下</string>
7       <string name="left">左</string>
8       <string name="right">右</string>
9       <string name="center">中</string>
10  </resources>
```

2.3 代码分析

2.3.1 线性布局

线性布局(LinearLayout)是 Android 中最基础也是最常用的布局管理器,所谓线性是指该布局中所有控件的排列都是按照同一方向,方向只有两种,即垂直和水平。当控件在某一方向上已经排满整个屏幕时,再添加其他控件时,该控件将无法显示,也就是说线性布局不会为了显示控件而改变方向。线性布局中常用的属性如下。

- android:orientation：设置线性布局的方向,该属性的值只有 vertical(垂直方向)和 horizontal(水平方向)两种,默认值为 horizontal。
- android:gravity：设置线性布局内控件的对齐方式,可以同时指定多种对齐方式的组合,多个属性之间用竖线隔开,但竖线前后不能出现空格。例如 bottom|center_horizontal 表示控件出现在屏幕底部,而且水平居中。

2.3.2 按比例分割屏幕

Android 平台以其开源、低成本、易开发等特性受到广大手机软件开发者、设备制造商等的青睐,市面上 Android 手机的尺寸、屏幕分辨率多种多样。为了让界面适合于不同屏幕尺寸的手机,通常会按比例分配控件的大小,而很少使用固定的像素。在 Android 中为控件提供了一个 android:layout_weight 属性,该属性表示的是控件所占屏幕剩余空间的权重,如果只有一个控件设置了该属性,则该控件会占满所有的剩余空间,如果有多个控件设置了该属性,则多个控件按照比例大小分配剩余空间。

例如有 3 个控件 a、b、c 水平排列,它们的宽度值都为 wrap_content 时,整个屏幕宽度为 X,它们的 android:layout_weight 属性值分别为 1、2、3,则表示第 1 个控件除了自己的宽度以外还额外占剩余空间宽度的 $1/6(6=1+2+3)$。以此类推,第 2 个控件还额外占剩余空间宽度的 $1/3$,第 3 个控件还额外占剩余空间宽度的 $1/2$。

剩余空间的宽度＝X－a 控件的宽度－b 控件的宽度－c 控件的宽度

那么,此时这 3 个控件的宽度之比为(a 控件的宽度＋$1/6×$剩余空间的宽度)∶(b 控件的宽度+1/3×剩余空间的宽度)∶(c控件的宽度+1/2×剩余空间的宽度),由于a、b、c控件自身的宽度关系不确定,因此它们之间的宽度比没有关系,达不到按比例分割的目的。要想按比例分割,此时有两种解决方案,一是设置所有控件的宽度为0;二是设置所有控件的宽度为match_parent。

当设置所有控件的宽度都为0时,此时剩余空间就是$X-0-0-0=X$,这样它们的宽度比就是$(0+1/6\times X)\colon(0+1/3\times X)\colon(0+1/2\times X)=1\colon2\colon3$。

当设置所有控件的宽度为match_parent时,此时剩余空间就是$X-X-X-X=-2X$,这样它们的宽度比就是$(X+1/6\times(-2X))\colon(X+1/3\times(-2X))\colon(X+1/2\times(-2X))=2\colon1\colon0$,即c控件无法显示。

通过上面的计算可知,如果想要控件的宽度成一定比例,最好的方式是设置它们的宽度为0,然后设置它们的android:layout_weight的值成一定比例。虽然通过设置它们的宽度为match_parent也能达到按比例分割的目的,但是计算起来比较麻烦。

因此在本例中,要想将整个屏幕按比例1∶4∶1分割成上、中、下3个部分,只需将3个部分控件的高度设为0,然后将其android:layout_weight的值设为1∶4∶1即可;要想将中间部分按比例1∶4∶1分割成左、中、右3个部分,只需将中间部分所包含的3个控件的宽度设为0,然后将其android:layout_weight的值设为1∶4∶1即可。

2.4 知识扩展

在实际应用时,大家经常会遇到一个屏幕难以显示完所有内容的情况,例如网页信息。如果确实需要显示超出屏幕的内容,可以在控件之外添加一个滚动条,由滚动条包裹该控件。滚动条有两种,即ScrollView(垂直滚动条)和HorizontalScrollView(水平滚动条),**在一个滚动条内部最多只能直接包裹一个控件**。例如,如果在水平线性布局中添加10个控件,按照屏幕的大小最多只能显示5个半控件,此时第6个控件由于有一部分可以显示,所以它会被压缩从而全部显示出来,而另外4个控件是完全不能显示的;如果需要显示出超出屏幕的那4个控件,可以为水平线性布局添加水平滚动条。垂直滚动条和水平滚动条可以混合使用,即可以在水平滚动条内部包裹一个垂直滚动条或者在垂直滚动条内部包裹一个水平滚动条。例如,显示大图片时,可以添加水平滚动条和垂直滚动条,从而查看图片全部。

2.5 思考与练习

(1) 层布局(FrameLayout)中的控件能否使用android:layout_weight属性,如果不能,请说明原因。

(2) 在下列关于线性布局的描述中,正确的是(   )。

    A) 水平线性布局中所有的控件都是按照水平方向一个挨着一个排列的,超出屏幕的宽度后,将会自动生成水平滚动条,拖动滚动条可查看其他控件

    B) 水平线性布局中所有的控件都是按照水平方向一个挨着一个排列的,超出屏

幕的宽度后,将会自动换行显示其他控件

C) 水平线性布局中所有的控件都是按照水平方向一个挨着一个排列的,超出屏幕的宽度后,将不会显示多余的控件

D) 水平线性布局中所有的控件都是按照水平方向一个挨着一个排列的,超出屏幕的宽度后,再添加控件,程序运行时将报错

(3) 在水平线性布局中,为了使控件的宽度成一定比例,需要使用(　　)属性。

A) android:layout_width　　　　B) android:layout_weight

C) android:layout_margin　　　　D) android:layout_gravity

(4) 在 ScrollView 垂直滚动条中,最多可直接包含(　　)个子控件。

A) 0　　　　B) 1　　　　C) 2　　　　D) 无数

第 3 章

我的课表——表格布局的应用

3.1 案例概述

本案例主要介绍表格布局(TableLayout)的使用,通过表格布局能够很方便地实现控件宽度一致、上下对齐的效果,对于一些比较规则的界面(例如几行几列)很实用。本案例实现一个简单的展示课表的功能,对于课表而言,每天课程的节数是相同的,只是上课的内容不同,是典型的几行几列表格。单击中间空白的按钮可以设置课程,用户也可以对已有课程进行修改。本案例的程序运行效果如图 3-1 所示。

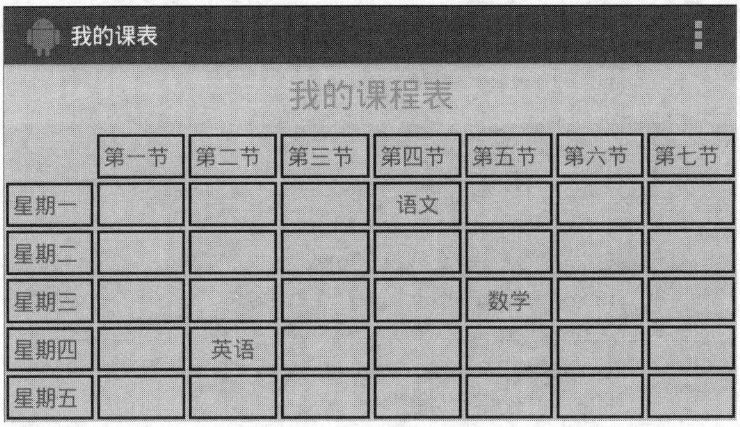

图 3-1 程序运行效果图

3.2 关键代码

布局文件:03\CourseList\res\layout\activity_main.xml

```
1   <TableLayout xmlns:android="http://schemas.android.com/apk/res/android"
2       android:layout_width="match_parent"
3       android:layout_height="match_parent"
4       android:background="#aabbcc">
5       <TextView
6           android:layout_width="match_parent"
7           android:layout_height="wrap_content"
```

```
 8            android:gravity="center"
 9            android:text="@string/mycur"
10            android:textColor="#ff2233"
11            android:background="#ccbbaa"
12            android:padding="5dp"
13            android:layout_marginBottom="5dp"
14            android:textSize="24sp"/>
15     <TableRow>
16         <TextView
17             style="@style/textView"
18             android:background="#00000000"/>
19         <TextView
20             style="@style/textView"
21             android:layout_column="1"
22             android:text="@string/first"/>
23         <TextView
24             style="@style/textView"
25             android:text="@string/second"/>
26         <TextView
27             style="@style/textView"
28             android:text="@string/third"/>
29         <TextView
30             style="@style/textView"
31             android:text="@string/forth"/>
32         <TextView
33             style="@style/textView"
34             android:text="@string/fifth"/>
35         <TextView
36             style="@style/textView"
37             android:text="@string/sixth"/>
38         <TextView
39             style="@style/textView"
40             android:text="@string/seventh"/>
41     </TableRow>
42     <TableRow>
43         <TextView
44             style="@style/textView"
45             android:text="@string/mon"/>
46         <Button style="@style/btn"/>
47         <Button style="@style/btn"/>
48         <Button style="@style/btn"/>
49         <Button style="@style/btn"/>
50         <Button style="@style/btn"/>
51         <Button style="@style/btn"/>
```

```
52        <Button style="@style/btn"/>
53      </TableRow>
54      <TableRow>
55        <TextView
56          style="@style/textView"
57          android:text="@string/tue"/>
58        <Button style="@style/btn"/>
59        <Button style="@style/btn"/>
60        <Button style="@style/btn"/>
61        <Button style="@style/btn"/>
62        <Button style="@style/btn"/>
63        <Button style="@style/btn"/>
64        <Button style="@style/btn"/>
65      </TableRow>
66      <TableRow>
67        <TextView
68          style="@style/textView"
69          android:text="@string/wed"/>
70        <Button style="@style/btn"/>
71        <Button style="@style/btn"/>
72        <Button style="@style/btn"/>
73        <Button style="@style/btn"/>
74        <Button style="@style/btn"/>
75        <Button style="@style/btn"/>
76        <Button style="@style/btn"/>
77      </TableRow>
78      <TableRow>
79        <TextView
80          style="@style/textView"
81          android:text="@string/thu"/>
82        <Button style="@style/btn"/>
83        <Button style="@style/btn"/>
84        <Button style="@style/btn"/>
85        <Button style="@style/btn"/>
86        <Button style="@style/btn"/>
87        <Button style="@style/btn"/>
88        <Button style="@style/btn"/>
89      </TableRow>
90      <TableRow>
91        <TextView
92          style="@style/textView"
93          android:text="@string/fri"/>
94        <Button style="@style/btn"/>
95        <Button style="@style/btn"/>
```

```
96          <Button style="@style/btn"/>
97          <Button style="@style/btn"/>
98          <Button style="@style/btn"/>
99          <Button style="@style/btn"/>
100         <Button style="@style/btn"/>
101     </TableRow>
102 </TableLayout>
```

字符串常量文件：03\CourseList\res\values\strings.xml

```
1  <?xml version="1.0" encoding="utf-8"?>
2  <resources>
3      <string name="app_name">我的课表</string>
4      <string name="action_settings">Settings</string>
5      <string name="mon">星期一</string>
6      <string name="tue">星期二</string>
7      <string name="wed">星期三</string>
8      <string name="thu">星期四</string>
9      <string name="fri">星期五</string>
10     <string name="first">第一节</string>
11     <string name="second">第二节</string>
12     <string name="third">第三节</string>
13     <string name="forth">第四节</string>
14     <string name="fifth">第五节</string>
15     <string name="sixth">第六节</string>
16     <string name="seventh">第七节</string>
17     <string name="mycur">我的课程表</string>
18 </resources>
```

样式文件：03\CourseList\res\values\styles.xml

```
1  <resources xmlns:android="http://schemas.android.com/apk/res/android">
2      <style name="AppBaseTheme" parent="android:Theme.Light"></style>
3      <style name="AppTheme" parent="AppBaseTheme"></style>
4      <style name="textView">
5          <item name="android:layout_width">0dp</item>
6          <item name="android:layout_weight">1</item>
7          <item name="android:layout_height">wrap_content</item>
8          <item name="android:layout_margin">1dp</item>
9          <item name="android:textSize">16sp</item>
10         <item name="android:background">@drawable/bg</item>
11     </style>
12     <style name="btn">
13         <item name="android:layout_width">0dp</item>
```

```
14      <item name="android:layout_weight">1</item>
15      <item name="android:layout_height">wrap_content</item>
16      <item name="android:layout_margin">1dp</item>
17      <item name="android:textSize">16sp</item>
18      <item name="android:background">@drawable/bg</item>
19      <item name="android:minHeight">0dp</item>
20      <item name="android:minWidth">0dp</item>
21      <item name="android:textColor">#0000ff</item>
22      <item name="android:onClick">setCourse</item>
23    </style>
24  </resources>
```

程序代码：03\CourseList\src\iet\jxufe\cn\android\courselist\MainActivity.java

```java
1   public class MainActivity extends Activity {
2       private Button btn;                                    //记录单击的按钮
3       @Override
4       protected void onCreate(Bundle savedInstanceState){
5           super.onCreate(savedInstanceState);
6           setContentView(R.layout.activity_main);
7       }
8       public void setCourse(View view){                      //单击按钮的事件处理
9           btn=(Button)view;                                  //获取被单击的按钮
10          Builder courseBuilder=new AlertDialog.Builder(this);   //创建对话框
11          courseBuilder.setTitle("请输入课程名");             //设置对话框标题
12          final EditText editText=new EditText(this);        //创建一个文本编辑框
13          editText.setText(btn.getText().toString());//设置文本编辑框内容
14          courseBuilder.setView(editText);                   //将文本编辑框添加到对话框中
15          courseBuilder.setPositiveButton("确定",new OnClickListener(){
16              @Override
17              public void onClick(DialogInterface dialog, int which){
18                  btn.setText(editText.getText().toString());        //设置课程
19              }
20          });                                                //为对话框添加"确定"按钮
21          courseBuilder.create().show();                     //创建并显示对话框
22      }
23      @Override
24      public boolean onCreateOptionsMenu(Menu menu){
25          //Inflate the menu; this adds items to the action bar if it is present.
26          getMenuInflater().inflate(R.menu.main, menu);
27          return true;
28      }
29  }
```

3.3 代码分析

3.3.1 界面分析

课表界面分析如图 3-2 所示,整体采用表格布局(TableLayout),其中,标题文本单独占一行,并居中显示,下面是规则的 6 行 8 列的表格,只不过第 1 行第 1 列为空白。在这些单元格中,第 1 行和第 1 列所有的单元格都是 TextView,用于显示星期和上课时间的提示,其他单元格都是按钮,用于显示具体的课程,单击按钮可以对课程进行设置和修改。表格中每个单元格的大小、宽度、高度等都一致,并且每个单元格都包含有边框,为了复用代码,可以单独为其定义样式。在此定义两个样式,一个是针对 TextView 的,一个是针对 Button 的。

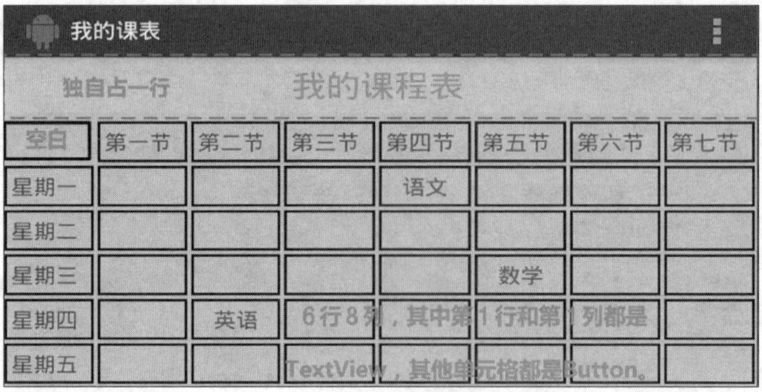

图 3-2 课表界面分析图

3.3.2 表格布局

表格布局(TableLayout)以行和列的形式来管理界面控件的布局管理器,但在表格布局中并不能明确声明包含几行几列。可通过添加 TableRow 来增加行,TableRow 本身也是容器,可以在 TableRow 中继续添加控件,每添加一个控件表示在该行中增加一列。如果直接在表格布局中添加控件,则该控件将会单独占一行。在表格布局中,每列的宽度都是一样的,列的宽度由该列中最宽的那个单元决定,整个表格布局的宽度则取决于表格里的内容,但不能超过父容器的宽度。表格布局中常见的属性如下。

- android:collapseColumns:隐藏指定的列,其值为列所在的序号,从 0 开始,如果需要隐藏多列,可用逗号隔开这些序号。隐藏该列后,后面的列会占用该列的位置。
- android:shrinkColumns:压缩指定的列,从而使整行能够完全显示,不会超出屏幕。该属性用于某一行的内容超过屏幕的宽度时,此时,会使指定列压缩换行,其值为列所在的序号。如果没有指定该属性,则超出屏幕的部分将自动截取,不会显示。

- android:stretchColumns：扩展指定的列,以填充屏幕中的空白部分。该属性用于行的内容不足以填充整个屏幕时,此时,指定列的宽度将变大以填满空白部分,其他列的宽度不变。表格中可以包含多行,每行的列数可能不相同,但同一列的宽度是一致的,并不会因为某一行有空余而使该行中某一列的宽度变大,也不会因为其他行没有空余而使它们的该列宽度不变。
- android:layout_column：指定控件在 TableRow 中所处的列。如果没有设置该属性,在默认情况下,控件在一行中是一列挨着一列排列的。通过设置该属性,可以指定控件所在的列,从而达到中间某一个列为空的效果。
- android:layout_span：指定某一控件所跨越的列数,即将多列合并为一列。

要想实现课表界面中的标题独占一行的效果,只需将标题 TextView 直接放入 TableLayout 中即可。

要想实现课表界面中第 1 行第 1 列空白的效果,只需设置第 1 行中第 1 个控件的 android:layout_column 值为 1 或者将第 1 行中第 1 个控件的背景设置为透明即可。

3.3.3 为 TextView 添加边框

在 Android 中,TextView 控件并不存在设置边框的属性,那么怎样为 TextView 添加边框呢?主要有 3 种方法,其一,自定义一个控件,该控件继承 TextView,然后重写其 onDraw()方法,绘制边框;其二,设计一张背景图片,该图片透明且带有边框;其三,定义一个 shape 类型的 XML 文件,然后将其作为 TextView 的背景。

在这里采用第 3 种方式,即自定义一个矩形,矩形的边框为 2dp,颜色为黑色,而矩形的中间填充部分为透明,然后将该矩形作为 TextView 的背景。具体代码如下:

自定义图形文件:03\CourseList\res\drawable-hdpi\bg.xml

```
1  <?xml version="1.0" encoding="utf-8"?>
2  <shape xmlns:android="http://schemas.android.com/apk/res/android"
3      android:shape="rectangle" >
4      <solid android:color="#00000000" /><!--矩形颜色为透明 -->
5      <padding android:left="5dp"
6          android:top="5dp"
7          android:right="5dp"
8          android:bottom="5dp"/>
9      <stroke
10         android:width="2dp"
11         android:color="#000000" /><!--矩形边框为两个像素,颜色为黑色 -->
12 </shape>
```

3.3.4 定义样式

在课表界面中,TextView 和 Button 控件比较多,每个控件都需要设置宽度、高度、文字大小、背景颜色等,并且大部分 TextView 和 Button 的这些属性的值都是一致的,分别设置代码比较冗长,特别是想对这些 TextView 换一种风格时需要一个个去修改,比较麻烦。在 Android 中,用户可以将控件的一些共同属性提取出来,定义成样式,当控件需要使用这些属性时,只需要调用该样式即可;当用户需要改变整体的风格时,只需要修改样式定义文件即可。这样只需要更改一处,就可以达到所有控件全部变化的目的,非常方便。如果用户只需要更改某一个控件的风格,可以直接在该控件中对属性重新赋值,此时它会覆盖样式中的属性。

在 Android 中定义样式文件,通常是在 res\values\styles.xml 文件中的<resource>根元素下添加<style>元素,它的 name 属性用来指定样式名。样式中的每一个属性值都用<item>标签表示,<item>标签的 name 属性指定具体的属性,<item>标签的内容为属性的值。在引用时,只需将组件的 style 属性值设置为**@style/样式名**即可,具体代码见本案例中提供的 styles.xml 文件。

3.3.5 直接绑定到标签

Android 中的事件处理有 3 种方式,即基于监听的事件处理机制、基于回调的事件处理机制以及直接绑定到标签。其中,基于监听的事件处理机制最灵活,使用最广泛;基于回调的事件处理机制仅限于系统提供了一些固定的事件处理,如触摸事件、按键事件等,重写系统提供的一些回调方法即可;直接绑定到标签仅限于个别事件,如单击事件,但非常简单、方便。

在本案例中包含 35 个按钮,单击任何一个按钮都需要进行事件处理,如果采用基于监听的事件处理需要为这些按钮分别添加 android:id 属性,然后根据 ID 找到相应按钮,再为这些按钮注册单击事件处理器,非常烦琐。在此,可采用直接绑定到标签的方式,为按钮标签添加 onClick 属性,由于所有的按钮都需要,可以将该属性添加到样式文件中。该属性值即具体处理单击事件的方法,在此为 setCourse。因此,需要在 Activity 中添加一个 public void setCourse(View view)方法,除了方法名之外其他都是固定写法,其中传递的参数 View 为具体被单击的按钮。

3.4 知识扩展

在 Android 中,手机屏幕通常是竖屏的,然而在某些情况下,界面横屏显示会更加美观,最常见的就是视频播放,一般的视频都是 16:9 或者 4:3,竖屏播放比较小,不太美观。在 Android 中为开发者提供了更改界面为横屏显示的方法,有以下两种:

(1) 在 AndroidManifest.xml 清单文件中对需要横屏显示的 Activity 添加 android:screenOrientation="landscape"属性。

(2) 在代码中进行判断,如果是竖屏,则将其设置为横屏。代码如下:

```
1   if(getRequestedOrientation()!=ActivityInfo.SCREEN_ORIENTATION_LANDSCAPE){
2       setRequestedOrientation(ActivityInfo.SCREEN_ORIENTATION_LANDSCAPE);
3   }
```

3.5 思考与练习

(1) 本案例中表格布局刚刚填满了整个屏幕,如果现在每天晚上要上晚自习,即一天有10节课,那么应该如何变化使得可以查看10节课?(提示:添加滚动条)

(2) 在下面自定义 style 的方式中,正确的是( )。

A) ```
<resources>
    <style name="myStyle">
        <item name="android:layout_width">match_parent</item>
    </style>
</resources>
```

B) ```
<style name="myStyle">
    <item name="android:layout_width">match_parent </item>
</style>
```

C) ```
<resources>
    <item name="android:layout_width">match_parent </item>
</resources>
```

D) ```
<resources>
    <style name="android:layout_width">match_parent </style>
</resources>
```

(3) 在下列关于表格布局的描述中,不正确的是( )。

A) 表格布局从线性布局继承而来

B) 表格布局中可明确指定包含多少行多少列

C) 在表格布局中可设置某一控件占多列

D) 如果直接向表格布局中添加控件,而不是在 TableRow 中添加,则该控件将单独占一行

(4) 在表格布局中,设置某一列为可压缩列的正确方法是( )。

A) 设置 TableLayout 的属性:android:stretchColumns="x",x 表示列的序号

B) 设置 TableLayout 的属性:android:shrinkColumns="x",x 表示列的序号

C) 设置具体列的属性:android:stretchable="true"

D) 设置具体列的属性:android:shrinkable="true"

(5) 运用表格布局设置3行3列的按钮,要求第1行中有一列空着,第3列被拉伸,第3行中有一个按钮占两列,运行效果如图3-3所示。

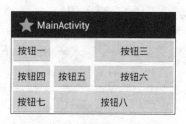

图 3-3 习题 5 运行效果图

第4章

闪烁霓虹灯——层布局的应用

4.1 案例概述

本案例主要介绍层布局的使用，通过层布局可以很方便地实现多个控件叠加的效果。本案例结合定时器和 Handler 消息传递机制实现闪烁霓虹灯的效果，通过 Timer 定时发送消息，Handler 在接收消息后进行相关处理，改变控件的背景颜色实现闪烁的效果。本案例的程序运行效果如图 4-1 和图 4-2 所示。

图 4-1 程序运行某一时刻图 1

图 4-2 程序运行某一时刻图 2

4.2 关键代码

布局文件：04\FrameLayoutTest\res\layout\activity_main.xml

```
1   <FrameLayout xmlns:android="http://schemas.android.com/apk/res/android"
2       xmlns:tools="http://schemas.android.com/tools"
3       android:layout_width="match_parent"
4       android:layout_height="match_parent"
5       android:background="#aabbcc">
```

```
 6      <TextView
 7          android:id="@+id/text01"
 8          android:layout_width="240dp"
 9          android:layout_height="240dp"
10          android:layout_gravity="center"/>
11      <TextView
12          android:id="@+id/text02"
13          android:layout_width="200dp"
14          android:layout_height="200dp"
15          android:layout_gravity="center"/>
16      <TextView
17          android:id="@+id/text03"
18          android:layout_width="160dp"
19          android:layout_height="160dp"
20          android:layout_gravity="center"/>
21      <TextView
22          android:id="@+id/text04"
23          android:layout_width="120dp"
24          android:layout_height="120dp"
25          android:layout_gravity="center"/>
26      <TextView
27          android:id="@+id/text05"
28          android:layout_width="80dp"
29          android:layout_height="80dp"
30          android:layout_gravity="center"/>
31      <ImageView
32          android:src="@drawable/ic_launcher"
33          android:layout_width="wrap_content"
34          android:layout_height="wrap_content"
35          android:layout_gravity="center"
36          android:contentDescription="@string/imgInfo"/>
37  </FrameLayout>
```

由于字符串常量文件中的内容比较简单，在此不再列出。

程序代码：04\FrameLayoutTest\src\iet\jxufe\cn\android\framelayouttest\MainActivity.java

```
1  public class MainActivity extends Activity {
2      private int[]textIds=new int[] { R.id.text01,R.id.text02,R.id.text03,
          R.id.text04,R.id.text05 };//定义一个数组,用于存储所有 TextView 的 ID
3      private int[]colors=new int[] { Color.RED,Color.MAGENTA,Color.GREEN,
          Color.YELLOW,Color.BLUE };      //定义一个数组,用于存储 5 种颜色
4      //定义一个数组,数组元素为 TextView,数组的长度由前面的数组决定
       private TextView[] views=new TextView[textIds.length];
5      private Handler mHandler;
```

```
6        private int current=0;          //记录从哪个颜色开始
7        protected void onCreate(Bundle savedInstanceState){
8            super.onCreate(savedInstanceState);
9            setContentView(R.layout.activity_main);
10           for(int i=0;i<textIds.length;i++){
11               views[i]=(TextView)findViewById(textIds[i]);
12           }//循环遍历 ID 数组,根据 ID 获取控件,然后将控件赋给 TextView 数组中的元素
13           //创建 Handler 对象,用于接收消息并处理
14           mHandler=new Handler(){
15               public void handleMessage(Message msg){    //处理消息的方法
16                   if(msg.what==0x11){       //判断消息是否为指定的消息
17                       //循环设置 TextView 的背景颜色
18                       for(int i=0;i<views.length;i++){
19                           views[i].setBackgroundColor(colors[(i+current)
20                               % colors.length]);
21                       }
22                       current=(current+1)% colors.length;
                         //使开始颜色的序号加 1,如果已经是最后一个,则从第一个开始
23                   }
24               }
25           };
26           Timer timer=new Timer();     //创建定时器对象
27           timer.schedule(new TimerTask(){
28               public void run(){
29                   mHandler.sendEmptyMessage(0x11);
30               }
31           },0,3000);                    //开启定时任务,每隔 3000ms 发送一次消息
32       }
33       @Override
34       public boolean onCreateOptionsMenu(Menu menu){
35           getMenuInflater().inflate(R.menu.main,menu);
36           return true;
37       }
38   }
```

4.3 代码分析

4.3.1 界面分析

该界面中包含 6 个控件,即一个 ImageView 控件、5 个 TextView 控件。其中,ImageView 控件位于屏幕的正中央,显示的图片为应用的图标;5 个 TextView 控件也都位于屏幕的正中央,最底层的 TextView 控件最大,其他控件依次减小。由于这些 TextView 控件的背景颜色需要动态变化,这是通过 XML 布局文件无法实现的,因此需要为这些

TextView 控件添加 ID 属性,然后在 Java 代码中就可以根据 ID 找到相应的控件。

根据界面的特点,存在相互叠加的效果,本案例无法通过线性布局或者表格布局来实现。在此介绍两种新的布局,即相对布局和层布局。

4.3.2 相对布局

相对布局(RelativeLayout)顾名思义是指布局里控件的位置相对于某个已有控件的位置应该如何摆放。这种布局的关键是找到一个合适的参照物,如果甲组件的位置需要根据乙组件的位置来确定,那么要求先定义乙组件,再定义甲组件。

在相对布局中,参照物通常有两种,一种是父容器,即当前的相对布局;一种是某个已有的控件。由于父容器有且只有一个,所以当控件的位置是相对于父容器的方位或对齐关系时,取值只能为 true 或 false;当参照物是某个已有的控件时,则需要指明该控件的 ID。相对布局中主要的属性如表 4-1 所示。

表 4-1 相对布局中的常用属性

属　　性	作　　用
android:layout_centerHorizontal	设置该组件是否位于父容器的水平居中位置
android:layout_centerVertical	设置该组件是否位于父容器的垂直居中位置
android:layout_centerInParent	设置该组件是否位于父容器的正中央位置
android:layout_alignParentTop	设置该组件是否与父容器的顶端对齐
android:layout_alignParentBottom	设置该组件是否与父容器的底端对齐
android:layout_ alignParentLeft	设置该组件是否与父容器的左边对齐
android:layout_ alignParentRight	设置该组件是否与父容器的右边对齐
android:layout_toRightOf	指定该组件位于给定的 ID 控件的右侧
android:layout_toLeftOf	指定该组件位于给定的 ID 控件的左侧
android:layout_above	指定该组件位于给定的 ID 控件的上方
android:layout_below	指定该组件位于给定的 ID 控件的下方
android:layout_alignTop	指定该组件与给定的 ID 控件的上边界对齐
android:layout_ alignBottom	指定该组件与给定的 ID 控件的下边界对齐
android:layout_ alignLeft	指定该组件与给定的 ID 控件的左边界对齐
android:layout_ alignRight	指定该组件与给定的 ID 控件的右边界对齐

通常,一个控件的位置的确定需要两个方面的信息,即方位以及对齐方式。

注意:相对布局中的参照物只能是相对布局内部的控件,不能是相对布局外面的控件。

4.3.3 层布局

层布局或者帧布局(FrameLayout)是指布局内的每个控件单独占一帧或一层,该层中未包含内容的部分将是透明的。控件添加的顺序即层叠加的次序,后面添加的控件会覆盖前面的控件,如果后添加的控件未能完全覆盖前面的控件,则未覆盖的部分将会显示。层布局中控件的位置可通过 android:layout_gravity 属性进行设置。通过层布局能够很方便地实现多个控件叠加或者渐变的效果。

注意:由于层布局中的每个控件单独占一层,也就是说层布局中各个控件之间不存在任何关系,即不能在层布局中按比例分割屏幕。

4.3.4 定时器

在本案例中需要实现周期性改变控件背景的效果,可以通过启动一个线程,然后在线程体中执行死循环,每执行一次循环,在循环体中让线程休眠 3000ms,从而达到每隔 3000ms 变换一次背景的目的。在 Android 中对此有更好的封装,提供了定时器类 Timer,当需要执行周期性的操作时,只需要创建 Timer 对象,然后调用它的 schedule() 方法即可。该方法可以传递 3 个参数,第一个参数为 TimerTask 对象,表示具体要执行的操作,TimerTask 类自身是一个抽象类,包含一个抽象方法 run(),自己不能实例化,必须创建一个该类的子类实现其 run() 方法;第二个参数为执行该操作延迟的时间,单位为毫秒;第 3 个参数为执行该操作的周期,单位为毫秒。程序代码的第 27~31 行即表示每隔 3000ms 发送一次消息。

为什么不直接在 run() 方法中执行更改控件背景的操作,而要发送消息,然后由 Handler 来接收和处理消息呢?这是因为在 Android 中界面控件是非线程安全的,所谓的非线程安全是指当多个线程对其进行操作时结果可能会不一致。为了避免出现这种情况,Android 中明确规定,所有对界面的操作只能放在主线程中,而不能在子线程中进行操作。虽然在程序中我们并未看到有子线程,但是 Timer 类实际上创建了一个子线程,run() 方法的执行不是在主线程中,所以不能在该方法中更改控件的背景。

4.3.5 Handler 消息传递

主线程能够对界面进行变化,但并不清楚应该什么时候去变化,子线程虽然想对界面进行变化,但自身又不能对其改变,这样就陷入一种矛盾之中,这时候需要借助一定的中介使得二者进行交互,Android 中的 Handler 消息传递机制就应运而生了。Handler 类的主要方法如表 4-2 所示。

表 4-2 Handler 类的主要方法

方法签名	描述
public void handleMessage(Message msg)	通过该方法获取并处理信息
public final boolean sendEmptyMessage(int what)	发送一个只含有标记的消息
public final boolean sendMessage(Message msg)	发送一个具体的消息
public final boolean hasMessages(int what)	监测消息队列中是否有指定标记的消息
public final boolean post(Runnable r)	将一个线程添加到消息队列中

从以上方法可以看出，Handler 类主要用于发送消息、接收和处理消息。执行过程为在子线程中当需要对界面进行操作时通过 Handler 发送消息，消息一旦发送成功，将会回调 Handler 类的 handleMessage(Message msg)方法，由于该方法在主线程中，因此能够对界面执行更改操作。Handler 消息传递机制可以归纳为谁发送谁处理，需要时发送消息，消息处理自动执行。

由于处理消息的 handleMessage(Message msg)方法是一种回调方法，当 Handler 接收到消息时由系统自动调用，因此，通常创建 Handler 对象时需要重写该方法。例如，程序代码的第 14～25 行，在该方法中写相关的业务逻辑。由于一个 Handler 对象可以发送多个消息，因此接收时要判断消息的类别，然后针对不同的消息做不同的处理。在 Handler 消息传递过程中，发送和处理消息的是同一个 Handler 对象，谁发送谁处理，由于需要在子线程中访问 Handler 对象，因此 Handler 对象必须是成员变量或者用 final 修饰的变量。

Message(消息)类用于封装消息的信息，包括消息的标记、内容等，主要有以下几个关键字段。

- what：消息的标记，int 类型，该标记是由用户自定义的，以便接收者确定该消息是什么，从而做出相应的处理。
- arg1 和 arg2：整型参数，这两个字段主要用于存放简单的整型数值，如果想存放复杂的数据，可通过 Message 对象的 setData()方法进行存放。
- obj：对象，Object 类型，传递给接收者的任意类型的对象。
- replyTo：Messenger 类型（信使），可选字段，用于答复该消息能够被发送，具体如何用取决于发送者和接收者。

Message(消息)类中的主要方法如表 4-3 所示。

表 4-3 Message 类中的主要方法

方 法 签 名	描　　述
public Message()	构造方法，推荐使用 Message.Obtain()
public void copyFrom(Message o)	复制指定消息的内容
public long getWhen()	获取消息发送的时间，单位为毫秒
public Bundle getData()	获取消息中的数据
public static Message obtain()	从消息池中获取一个消息
public void setData(Bundle data)	向消息中写入数据
public void setTarget(Handler target)	设置消息的目标对象

4.4 知识扩展

当需要在 Java 代码中动态改变界面中控件的状态信息时，一般需要在界面中为该控件添加 ID 属性，然后在代码中通过 findViewById()方法获取该控件，再调用相应方法进行改变。如果界面中存在多个需要改变的控件，则代码相对来说比较冗长，这时可根据实际情况进行简化。例如，在本案例中，界面中存在 5 个 TextView 控件，对这些控件的操作非常类似，都是更改背景颜色，因此可以将这些控件和控件的 ID 分别放在一个数组中，然后通过 for 循环依次获取 ID，在循环体中根据 ID 找到相关控件，再将其赋给控件数组中的相应控件。当需要对所有控件的背景颜色进行变化时，只需要进行一次 for 循环即可。

使用该方法精简代码时需要满足一定的条件：①控件的类型相同，并且数目比较多；②对所有控件的操作类似；③控件的操作满足一定的规律。

4.5　思考与练习

（1）由于本案例比较简单，实际上用相对布局也可以实现其效果，请尝试用相对布局实现。

（2）本案例使用 Timer 定时器实现周期性改变控件背景的功能，请尝试使用普通的线程实现。

（3）在相对布局中，如果想让一个控件居中显示，则可设置该控件的（　　）。

　　A）android:gravity="center"

　　B）android:layout_gravity="center"

　　C）android:layout_centerInParent="true"

　　D）android:scaleType="center"

（4）在相对布局中，下列属性值只能为 true 或 false 的是（　　）。

　　A）android:layout_alignTop

　　B）android:layout_alignParentTop

　　C）android:layout_toLeftOf

　　D）android:layout_above

第5章 简易计算器——布局的综合应用

5.1 案例概述

本案例设计和实现一个简单的计算器,使用不同的方式实现,既可以通过综合运用多种布局(例如线性布局、表格布局、相对布局)来实现,也可以通过 Android 4.0 新增的网格布局来实现。本案例的程序运行效果如图 5-1 所示,程序运行界面分析图如图 5-2 所示。

图 5-1 程序运行效果图   图 5-2 程序运行界面分析图

5.2 关键代码

布局文件:05\CalculateTest\res\layout\activity_main.xml

```
1    <TableLayout xmlns:android="http://schemas.android.com/apk/res/android"
2        xmlns:tools="http://schemas.android.com/tools"
3        android:layout_width="match_parent"
4        android:layout_height="match_parent"
5        android:padding="10dp"
```

```xml
 6      android:background="#aabbcc">
 7      <EditText
 8          android:layout_width="match_parent"
 9          android:layout_height="wrap_content"
10          android:layout_margin="10dp"
11          android:enabled="false"
12          android:gravity="right|bottom"
13          android:inputType="numberDecimal"
14          android:textSize="24sp"
15          android:textColor="#770000ff"
16          android:background="@drawable/border"
17          android:text="@string/zero"/>
18      <TableRow android:gravity="center_horizontal">
19          <Button
20              style="@style/myStyle"
21              android:text="@string/mc"/>
22          <Button
23              style="@style/myStyle"
24              android:text="@string/mr"/>
25          <Button
26              style="@style/myStyle"
27              android:text="@string/ms"/>
28          <Button
29              style="@style/myStyle"
30              android:text="@string/mplus"/>
31          <Button
32              style="@style/myStyle"
33              android:text="@string/mminus"/>
34      </TableRow>
35      <TableRow android:gravity="center_horizontal">
36          <Button
37              style="@style/myStyle"
38              android:text="@string/arrow"/>
39          <Button
40              style="@style/myStyle"
41              android:text="@string/ce"/>
42          <Button
43              style="@style/myStyle"
44              android:text="@string/c"/>
45          <Button
46              style="@style/myStyle"
47              android:text="@string/plusminus"/>
48          <Button
49              style="@style/myStyle"
```

```
50          android:text="@string/correct"/>
51      </TableRow>
52      <TableRow android:gravity="center_horizontal">
53          <Button
54              style="@style/myStyle"
55              android:text="@string/seven"/>
56          <Button
57              style="@style/myStyle"
58              android:text="@string/eight"/>
59          <Button
60              style="@style/myStyle"
61              android:text="@string/nine"/>
62          <Button
63              style="@style/myStyle"
64              android:text="@string/div"/>
65          <Button
66              style="@style/myStyle"
67              android:text="@string/mode"/>
68      </TableRow>
69      <TableRow android:gravity="center_horizontal">
70          <Button
71              style="@style/myStyle"
72              android:text="@string/four"/>
73          <Button
74              style="@style/myStyle"
75              android:text="@string/five"/>
76          <Button
77              style="@style/myStyle"
78              android:text="@string/six"/>
79          <Button
80              style="@style/myStyle"
81              android:text="@string/mul"/>
82          <Button
83              style="@style/myStyle"
84              android:text="@string/daoshu"/>
85      </TableRow>
86      <RelativeLayout
87          android:layout_width="wrap_content"
88          android:layout_height="wrap_content"
89          android:gravity="center_horizontal" >
90          <Button
91              android:id="@+id/one"
92              style="@style/myStyle"
93              android:layout_marginTop="2dp"
```

```xml
94              android:text="@string/one"/>
95          <Button
96              android:id="@+id/two"
97              style="@style/myStyle"
98              android:layout_alignTop="@id/one"
99              android:layout_toRightOf="@id/one"
100             android:text="@string/two"/>
101         <Button
102             android:id="@+id/three"
103             style="@style/myStyle"
104             android:layout_alignTop="@id/one"
105             android:layout_toRightOf="@id/two"
106             android:text="@string/three"/>
107         <Button
108             android:id="@+id/minus"
109             style="@style/myStyle"
110             android:layout_alignTop="@id/one"
111             android:layout_toRightOf="@id/three"
112             android:text="@string/minus"/>
113         <Button
114             android:id="@+id/equal"
115             style="@style/myStyle"
116             android:layout_height="100dp"
117             android:layout_alignTop="@id/one"
118             android:layout_toRightOf="@id/minus"
119             android:text="@string/equal"/>
120         <Button
121             android:id="@+id/plus"
122             style="@style/myStyle"
123             android:layout_alignBottom="@id/equal"
124             android:layout_toLeftOf="@id/equal"
125             android:text="@string/plus"/>
126         <Button
127             android:id="@+id/dot"
128             style="@style/myStyle"
129             android:layout_alignBottom="@id/equal"
130             android:layout_toLeftOf="@id/plus"
131             android:text="@string/dot"/>
132         <Button
133             style="@style/myStyle"
134             android:layout_width="120dp"
135             android:layout_alignBottom="@id/equal"
136             android:layout_toLeftOf="@id/dot"
137             android:text="@string/zero"/>
```

```
138         </RelativeLayout>
139 </TableLayout>
```

字符串常量文件：05\CalculateTest\res\values\strings.xml

```
1  <?xml version="1.0" encoding="utf-8"?>
2  <resources>
3      <string name="app_name">简易计算器</string>
4      <string name="action_settings">Settings</string>
5      <string name="mc">MC</string>
6      <string name="mr">MR</string>
7      <string name="ms">MS</string>
8      <string name="mplus">M+</string>
9      <string name="mminus">M-</string>
10     <string name="arrow">←</string>
11     <string name="ce">CE</string>
12     <string name="c">C</string>
13     <string name="plusminus">±</string>
14     <string name="correct">√</string>
15     <string name="zero">0</string>
16     <string name="one">1</string>
17     <string name="two">2</string>
18     <string name="three">3</string>
19     <string name="four">4</string>
20     <string name="five">5</string>
21     <string name="six">6</string>
22     <string name="seven">7</string>
23     <string name="eight">8</string>
24     <string name="nine">9</string>
25     <string name="plus">+</string>
26     <string name="minus">-</string>
27     <string name="mul">*</string>
28     <string name="div">/</string>
29     <string name="mode">%</string>
30     <string name="equal">=</string>
31     <string name="dot">.</string>
32     <string name="daoshu">1/x</string>
33 </resources>
```

在 CalculateTest\res\values\styles.xml 文件中的<resource>标签下添加如下代码

```
1  <style name=" myStyle">
2      <item name="android:layout_width">60dp</item>
3      <item name="android:layout_height">50dp</item>
```

```
    4        <item name="android:textSize">20sp</item>
    5        <item name="android:gravity">center</item>
    6    </style>
```

自定义边框：05\CalculateTest\res\drawable\border.xml

```
    1    <?xml version="1.0" encoding="utf-8"?>
    2    <shape xmlns:android="http://schemas.android.com/apk/res/android" >
    3        <padding
    4            android:bottom="5dp"
    5            android:left="5dp"
    6            android:right="5dp"
    7            android:top="5dp"/>
    8        <stroke
    9            android:width="1dp"
   10            android:color="#66666666"/>
   11    </shape>
```

5.3 代码分析

5.3.1 界面分析

该计算器界面包含一个文本编辑框和 28 个按钮，其中，文本编辑框的宽度填充整个手机屏幕，并且内容不允许用户直接编辑，而是根据用户的操作改变。在 28 个按钮中有 26 个按钮的样式是一样的，只有两个按钮比较特殊。其中，= 按钮的高度是普通按钮的两倍，0 按钮的宽度是普通按钮的两倍。

由于大部分按钮都是规则的，并且整体是按照行列的形式摆放的，因此可以考虑采用表格布局。然而，在表格布局中只允许控件跨列，不允许控件跨行，因此完全使用表格布局是无法实现该效果的，需要嵌套其他布局管理器。根据最后两行的特点，选择相对布局比较方便，因此在表格布局中嵌入一个相对布局即可实现计算器界面设计。在计算器界面中，按钮控件比较多，每个控件都需要设置宽度、高度、文字大小、对齐方式等，并且大部分按钮的这些属性值都是一致的，因此可参考前面介绍的方法减少代码量，具体代码见布局文件。

注意： 在使用表格布局时，如果控件没有放在 <TableRow> 标签内，该控件将单独占一行；在使用相对布局时，参考的对象一定是相对布局内部的控件或父容器。

5.3.2 网格布局

在 Android 4.0 中新增了一种布局管理器，即网格布局（GridLayout），该布局吸取了线性布局、表格布局、相对布局的一些优点。它把整个容器划分成 rows 行 columns 列个网格，每个网格可以放置一个控件，除此之外可以设置一个控件横跨多少列、纵跨多少行，以及控件的摆放方向是一行行排列，还是一列列摆放。网格布局中的主要属性如下。

- android:rowCount：设置该网格布局一共有多少行。
- android:columnCount：设置该网格布局一共有多少列。
- android:orientation：设置网格布局中控件的排列方向是水平还是垂直，默认是水平，即按行排列，如果不指定包含多少列，则网格布局只包含一行，类似于水平的线性布局；如果指定包含多少列，则会根据列来自动换行；如果是按列排列，则需要指定网格布局中包含多少行，否则只有一列，类似于垂直的线性布局；如果指定包含多少行，则会根据行来自动换列。
- android:layout_row：设置该控件所在网格行的序号。
- android:layout_rowSpan：设置该控件纵跨多少行。
- android:layout_column：设置该控件所在网格列的序号。
- android:layout_columnSpan：设置该控件横跨多少列。

使用网格布局实现计算器界面效果的代码如下。

布局文件：CalculateTest\res\layout\gridlayout.xml

```
1    <GridLayout xmlns:android="http://schemas.android.com/apk/res/android"
2        xmlns:tools="http://schemas.android.com/tools"
3        android:layout_width="match_parent"
4        android:layout_height="wrap_content"
5        android:orientation="horizontal"
6        android:padding="10dp"
7        android:columnCount="5">
8        <EditText
9            android:layout_height="wrap_content"
10           android:layout_columnSpan="5"
11           android:layout_margin="10dp"
12           android:enabled="false"
13           android:gravity="right|bottom"
14           android:inputType="numberDecimal"
15           android:textSize="24sp"
16           android:text="@string/zero"
17           android:textColor="#770000ff"
18           android:background="@drawable/border"
19           android:layout_gravity="fill_horizontal"/>
20       <Button
21           style="@style/myStyle"
22           android:text="@string/mc"/>
23       <Button
24           style="@style/myStyle"
25           android:text="@string/mr"/>
26       <Button
27           style="@style/myStyle"
28           android:text="@string/ms"/>
```

```xml
29    <Button
30        style="@style/myStyle"
31        android:text="@string/mplus"/>
32    <Button
33        style="@style/myStyle"
34        android:text="@string/minus"/>
35    <Button
36        style="@style/myStyle"
37        android:text="@string/arrow"/>
38    <Button
39        style="@style/myStyle"
40        android:text="@string/ce"/>
41    <Button
42        style="@style/myStyle"
43        android:text="@string/c"/>
44    <Button
45        style="@style/myStyle"
46        android:text="@string/plusminus"/>
47    <Button
48        style="@style/myStyle"
49        android:text="@string/correct"/>
50    <Button
51        style="@style/myStyle"
52        android:text="@string/seven"/>
53    <Button
54        style="@style/myStyle"
55        android:text="@string/eight"/>
56    <Button
57        style="@style/myStyle"
58        android:text="@string/nine"/>
59    <Button
60        style="@style/myStyle"
61        android:text="@string/div"/>
62    <Button
63        style="@style/myStyle"
64        android:text="@string/mode"/>
65    <Button
66        style="@style/myStyle"
67        android:text="@string/four"/>
68    <Button
69        style="@style/myStyle"
70        android:text="@string/five"/>
71    <Button
72        style="@style/myStyle"
```

```
73            android:text="@string/six"/>
74        <Button
75            style="@style/myStyle"
76            android:text="@string/mul"/>
77        <Button
78            style="@style/myStyle"
79            android:text="@string/daoshu"/>
80        <Button
81            style="@style/myStyle"
82            android:text="@string/one"/>
83        <Button
84            style="@style/myStyle"
85            android:text="@string/two"/>
86        <Button
87            style="@style/myStyle"
88            android:text="@string/three"/>
89        <Button
90            style="@style/myStyle"
91            android:text="@string/minus"/>
92        <Button
93            style="@style/myStyle"
94            android:text="@string/equal"
95            android:layout_rowSpan="2"
96            android:layout_gravity="fill_vertical"
97            />
98        <Button
99            style="@style/myStyle"
100            android:text="@string/zero"
101            android:layout_columnSpan="2"
102            android:layout_gravity="fill_horizontal"
103            />
104        <Button
105            style="@style/myStyle"
106            android:text="@string/dot"/>
107        <Button
108            style="@style/myStyle"
109            android:text="@string/plus"/>
110    </GridLayout>
```

在 MainActivity 中将"setContentView(R.layout.activity_main);"语句注释掉,添加语句"setContentView(R.layout.gridlayout);"。

注意:由于网格布局是 Android 4.0 中新增的布局,因此需要在 AndroidManifest.xml 文件中将最低版本设置为 14,即"android:minSdkVersion="14""。

5.4 知识扩展

在计算器界面设计完成后,单击按钮并没有任何反应,怎样才能实现具有运算功能的计算器呢?首先需要为这些按钮添加单击事件处理,为了区分这些按钮,需要为其添加 ID 属性,这样当用户单击某个按钮时,事件监听器能够捕获到用户的操作,然后进行相应的处理。计算器界面中的按钮可以大致分为两类,即数字按钮和运算符按钮。如果是数字按钮,则要将其转换成数字,并且可以多次单击数字按钮,将多个数字组成一个数,例如分别单击数字"1"、"4"、"6"按钮最后形成的数应该是 146,而不是用后面的取代前面的,结果为 6。如果是运算符按钮,则表明前一个数已经完成,接下来的数字形成一个新的数,当单击"="按钮时,在文本编辑框中显示表达式以及结果。本案例的运算演示效果如图 5-3 所示。

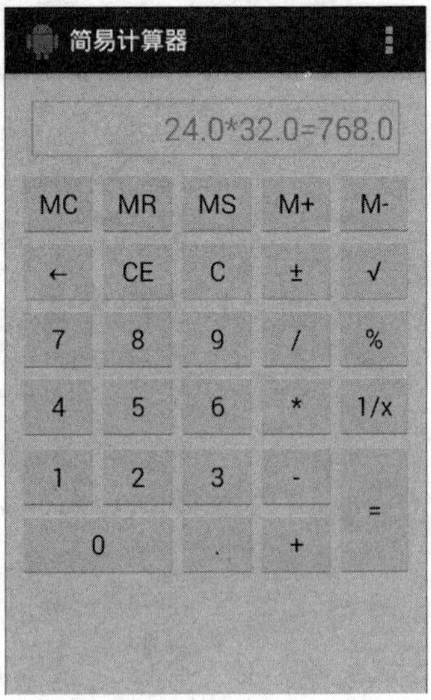

在本案例中,仅简单地实现加、减、乘、除、取余、退格和清空等功能,单击"←"按钮实现退格功能,将刚刚输入的数字消除;单击"C"按钮实现清空功能,将一切恢复原状。计算器的运算流程如图 5-4 所示。

图 5-3 运算演示效果

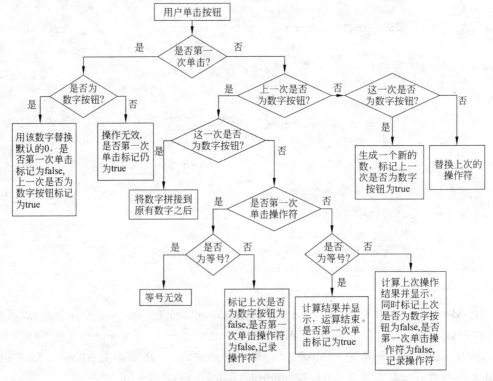

图 5-4 简易计算器的运算流程

第 5 章 简易计算器——布局的综合应用

此外,在执行计算的过程中还需要处理多次单击操作符、多次单击小数点、多次单击 0 等特殊情况。程序的关键代码如下。

程序代码：05\CalculateTest\src\iet\jxufe\cn\android\calculatetest\MainActivity.java

```
1   public class MainActivity extends Activity {
2       private int[]numberIds=new int[]{R.id.zero,R.id.one,R.id.two,R.id.three,
            R.id.four,R.id.five,R.id.six,R.id.seven,R.id.eight,R.id.nine,
            R.id.dot};                      //存放数字按钮和小数点按钮的 ID,共 11 个按钮
3       private int[]operationIds=new int[]{R.id.plus,R.id.minus,R.id.mul,
            R.id.div,R.id.mode,R.id.equal,R.id.clear,R.id.backspace};
            //存放操作符按钮的 ID,包括加、减、乘、除、取余、等于、退格、清除共 8 个按钮
4       private Button[]numberBtns=new Button[numberIds.length];
                                        //保存数字按钮的数组
5       private Button[]operationBtns=new Button[operationIds.length];
                                        //保存操作符按钮的数组
6       private EditText showInfo;          //显示信息的文本编辑框
7       private String str="0";             //保存文本编辑框中的值,默认为 0
8       private double num1;                //保存第一个操作数
9       private double num2;                //保存第二个操作数
10      private String operationStr="";     //保存操作符
11      private String result="";           //保存操作的表达式
12      private boolean isFirstClicked=true;    //是否第一次单击按钮
13      private boolean isLastNum=true;     //上一次输入的是否是数字,默认为 false
14      private boolean isOperationFirstClicked=true;   //默认是第一次单击操作符
15      @Override
16      protected void onCreate(Bundle savedInstanceState){
17          super.onCreate(savedInstanceState);
18          setContentView(R.layout.activity_main);  //综合使用表格布局与相对
                                                     //布局实现计算器界面
19          setContentView(R.layout.gridlayout);  //使用网格布局实现计算器界面
20          NumberBtnListener numberListener=new NumberBtnListener();
                                        //创建数字按钮单击监听器
21          for(int i=0;i<numberBtns.length;i++){  //循环为数组赋值,并为每一个
                                                    //数字按钮添加事件监听
22              numberBtns[i]=(Button)findViewById(numberIds[i]);
23              numberBtns[i].setOnClickListener(numberListener);
24          }
25          OperationBtnListener operationListener=new OperationBtnListener();
26          for(int i=0;i<operationBtns.length;i++){//循环为数组赋值,并为每一个
                                                    //操作按钮添加事件监听
27              operationBtns[i]=(Button)findViewById(operationIds[i]);
28              operationBtns[i].setOnClickListener(operationListener);
```

```
29        }
30        showInfo=(EditText)findViewById(R.id.showInfo);    //获取文本编辑框控件
31    }
32    private class NumberBtnListener implements OnClickListener {
                                              //监听数字按钮单击事件处理器
33        public void onClick(View v){
34            if(!isFirstClicked&&isLastNum){  //如果不是第一次单击按钮,并且
                                              //上一次输入也是数字
35                if(v.getId()==R.id.dot){
36                    if(str.indexOf(".")>=0){
37                        return;            //如果已经包含小数点,则不能再输入小数点
38                    }
39                }
40                str+=((Button)v).getText().toString();   //将数字拼接到原有
                                                          //数字之后
41                str=operationZero(str);    //执行去 0 操作
42            }else{
43                str=((Button)v).getText().toString();    //获取输入的数字
44                isFirstClicked=false;      //若是第一次单击为 false
45                isLastNum=true;            //若上一次输入的是数字为 true
46                if(v.getId()==R.id.dot){   //如果单击的是小数点
47                    str="0.";
48                }
49            }
50            showInfo.setText(str);         //显示输入的内容
51        }
52    }
53    private class OperationBtnListener implements OnClickListener {
                                              //监听操作符按钮单击事件处理器
54        public void onClick(View v){
55            if(isFirstClicked){    //如果是第一次单击按钮,并且单击的是操作符
56                showInfo.setText(str);
57                return;
58            }
59            if(v.getId()==R.id.clear){     //清空
60                clear();                   //执行清空操作
61                showInfo.setText(str);     //显示清空后的内容,即默认值
62                return;
63            }
64            if(v.getId()==R.id.backspace){ //退格
65                str=str.substring(0,str.length()-1); //截取从开始到倒数第
                                                     //二个字符的字符串
66                if(str.length()==0){       //如果当前的长度为 0,则取默认值 0
```

```
67                    str="0";
68                }
69                showInfo.setText(str);        //显示退格后的内容
70                return;
71            }
72            if(isLastNum){                    //如果上一次单击的是数字
73                if(isOperationFirstClicked){  //如果是第一次单击操作数
74                    if(!"=".equals(((Button)v).getText().toString())){
                                                  //如果不是等号
75                        num1=Double.parseDouble(str);
                                                  //第一个操作数
76                        operationStr=((Button)v).getText().toString();
                                                  //获取操作符
77                        isOperationFirstClicked=false; //若是第一次执行操
                                                  //作符为false
78                        isLastNum=false;        //若上一次单击的是数字为false
79                    }
80                }else{  //如果不是第一次单击操作数,则需要计算前面两个数的结果
81                    num2=Double.parseDouble(str);    //第二个操作数
82                    if("=".equals(((Button)v).getText().toString())){
                                                  //计算结果,运算结束
83                        double calReslt=getReslt(operationStr);
                                                  //获取运算结果
84                        showInfo.setText(result+"="+calResult+"");
                                                  //显示运算结果
85                        clear();                //执行清空操作
86                    }else{
87                        num1=getReslt(operationStr);//获取运算结果,并将其
                                                  //赋给num1
88                        showInfo.setText(result+"="+num1+"");
                                                  //显示运算结果
89                        operationStr=((Button)v).getText().toString();
                                                  //获取操作符
90                        isOperationFirstClicked=false; //第一次执行操作符
                                                  //为false
91                        isLastNum=false;        //若上一次单击的是数字为false
92                    }
93                }
94            }else{
95                operationStr=((Button)v).getText().toString();
                                                  //替换原来的操作符
96            }
97        }
```

```
98        }
99        private String operationZero(String str){              //对0的操作
100           if(str.indexOf(".")<0){                            //如果没有小数点,则不能以0开头
101               while(str.startsWith("0")&&str.length()>1){
102                   str=str.substring(1);                      //消除前面的0
103               }
104           }
105           return str;                                        //返回去0后的字符串
106       }
107       private double getReslt(String operationStr){
108           if("-".equals(operationStr)){                      //执行减法操作
109               result=num1+"-"+num2;                          //保存表达式
110               return num1-num2;                              //返回结果值
111           }else if("+".equals(operationStr)){                //执行加法操作
112               result=num1+"+"+num2;                          //保存表达式
113               return num1+num2;                              //返回结果值
114           }else if(" * ".equals(operationStr)){              //执行乘法操作
115               result=num1+" * "+num2;                        //保存表达式
116               return num1 * num2;                            //返回结果值
117           }else if("/".equals(operationStr)){                //执行除法操作
118               result=num1+"/"+num2;                          //保存表达式
119               return num1/num2;                              //返回结果值
120           }else if("%".equals(operationStr)){                //执行取余操作
121               result=num1+"%"+num2;                          //保存表达式
122               return num1%num2;                              //返回结果值
123           }
124           return num2;                                       //默认返回第二个操作数
125       }
126       public void clear(){                                   //清空,恢复原状
127           isFirstClicked=true;                               //第一次单击按钮为true
128           isOperationFirstClicked=true;                      //第一次单击运算符为true
129           num1=0;                                            //第一个操作数默认为0
130           num2=0;                                            //第一个操作数默认为0
131           operationStr="";                                   //运算符默认为空
132           str="0";                                           //文本编辑框默认显示0
133       }
134       @Override
135       public boolean onCreateOptionsMenu(Menu menu){
136           //Inflate the menu;this adds items to the action bar if it is present.
137           getMenuInflater().inflate(R.menu.main,menu);
138           return true;
139       }
140   }
```

5.5 思考与练习

（1）本案例通过综合使用表格布局和相对布局来实现计算器界面，能不能通过其他布局的组合方式来实现呢？请尝试实现。

（2）为了使每一行的内容水平居中，在每一个＜TableRow＞标签中都设置了 android:gravity＝"center_horizontal"，能不能通过在＜TableLayout＞标签中设置 android:gravity＝"center_horizontal"实现同样的效果？如果不能，请说明理由。

（3）在以下选项中，不属于 Android 布局管理器的是（　　）。
　　A）FrameLayout　　B）GridLayout　　C）BorderLayout　　D）TableLayout

第6章 页面滑动切换

6.1 案例概述

本案例主要实现页面滑动切换效果,通过 ViewPager 将前面所学的几个关键页面放在同一个应用中,滑动屏幕、单击标题的左边或者右边可以很方便地切换到上一个或下一个页面,同时下方表示页面序号的小图标也会发生变化,红色代表当前选中的页面,黄色代表未选中的页面,单击某个小图标时,也可以切换到相应的页面。本案例的程序运行效果如图 6-1 至图 6-3 所示。

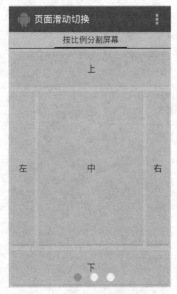

图 6-1 程序运行效果图 1

图 6-2 程序运行效果图 2

图 6-3 程序运行效果图 3

6.2 关键代码

布局文件:06\PageSwitcher\res\layout\activity_main.xml

```
1    <RelativeLayout xmlns:android="http://schemas.android.com/apk/res/android"
2        xmlns:tools="http://schemas.android.com/tools"
3        android:layout_width="match_parent"
```

```
4       android:layout_height="match_parent"
5       android:background="#aabbcc">
6       <android.support.v4.view.ViewPager
7           android:id="@+id/mViewPager"
8           android:layout_width="match_parent"
9           android:layout_height="match_parent">
10          <android.support.v4.view.PagerTabStrip
11              android:id="@+id/mTab"
12              android:layout_width="wrap_content"
13              android:layout_height="wrap_content"
14              android:layout_gravity="top"/>
15      </android.support.v4.view.ViewPager>
16      <LinearLayout
17          android:id="@+id/mImgs"
18          android:layout_width="wrap_content"
19          android:layout_height="wrap_content"
20          android:layout_alignParentBottom="true"
21          android:layout_centerHorizontal="true"
22          android:orientation="horizontal">
23      </LinearLayout>
24  </RelativeLayout>
```

按比例分割屏幕页面布局文件：06\PageSwitcher\res\layout\linearlayout.xml

```
1   <LinearLayout xmlns:android="http://schemas.android.com/apk/res/android"
2       xmlns:tools="http://schemas.android.com/tools"
3       android:layout_width="match_parent"
4       android:layout_height="match_parent"
5       android:orientation="vertical">
6       <Button
7           android:layout_width="match_parent"
8           android:layout_height="0dp"
9           android:layout_weight="1"
10          android:text="@string/up"/>
11      <LinearLayout
12          android:layout_width="match_parent"
13          android:layout_height="0dp"
14          android:layout_weight="4"
15          android:orientation="horizontal">
16          <Button
17              android:layout_width="0dp"
18              android:layout_height="match_parent"
19              android:layout_weight="1"
20              android:text="@string/left"/>
```

```
21      <Button
22          android:layout_width="0dp"
23          android:layout_height="match_parent"
24          android:layout_weight="4"
25          android:text="@string/center"/>
26      <Button
27          android:layout_width="0dp"
28          android:layout_height="match_parent"
29          android:layout_weight="1"
30          android:text="@string/right"/>
31      </LinearLayout>
32      <Button
33          android:layout_width="match_parent"
34          android:layout_height="0dp"
35          android:layout_weight="1"
36          android:text="@string/down"/>
37  </LinearLayout>
```

霓虹灯页面布局文件：06\PageSwitcher\res\layout\framelayout.xml

```
1   <FrameLayout xmlns:android="http://schemas.android.com/apk/res/android"
2       xmlns:tools="http://schemas.android.com/tools"
3       android:layout_width="match_parent"
4       android:layout_height="match_parent">
5       <TextView
6           android:id="@+id/text01"
7           android:layout_width="240dp"
8           android:layout_height="240dp"
9           android:background="#ff0000"
10          android:layout_gravity="center"/>
11      <TextView
12          android:id="@+id/text02"
13          android:layout_width="200dp"
14          android:layout_height="200dp"
15          android:background="#ffff00"
16          android:layout_gravity="center"/>
17      <TextView
18          android:id="@+id/text03"
19          android:layout_width="160dp"
20          android:layout_height="160dp"
21          android:background="#00ff00"
22          android:layout_gravity="center"/>
23      <TextView
24          android:id="@+id/text04"
```

```
25        android:layout_width="120dp"
26        android:layout_height="120dp"
27        android:background="#00ffff"
28        android:layout_gravity="center"/>
29    <TextView
30        android:id="@+id/text05"
31        android:layout_width="80dp"
32        android:layout_height="80dp"
33        android:background="#0000ff"
34        android:layout_gravity="center"/>
35    <ImageView
36        android:src="@drawable/ic_launcher"
37        android:layout_width="wrap_content"
38        android:layout_height="wrap_content"
39        android:layout_gravity="center"
40        android:contentDescription="@string/imageInfo"/>
41 </FrameLayout>
```

计算器页面布局文件：06\PageSwitcher\res\layout\gridlayout.xml

```
1  <GridLayout xmlns:android="http://schemas.android.com/apk/res/android"
2      xmlns:tools="http://schemas.android.com/tools"
3      android:layout_width="match_parent"
4      android:layout_height="match_parent"
5      android:orientation="horizontal"
6      android:padding="10dp"
7      android:columnCount="5">
8      <EditText
9          android:layout_height="60dp"
10         android:layout_columnSpan="5"
11         android:enabled="false"
12         android:gravity="right|bottom"
13         android:inputType="numberDecimal"
14         android:textSize="24sp"
15         android:text="@string/zero"
16         android:layout_gravity="fill_horizontal"/>
17     <Button
18         style="@style/myStyle"
19         android:text="@string/mc"/>
20     <Button
21         style="@style/myStyle"
22         android:text="@string/mr"/>
23     <Button
24         style="@style/myStyle"
```

```
25          android:text="@string/ms"/>
26      <Button
27          style="@style/myStyle"
28          android:text="@string/mplus"/>
29      <Button
30          style="@style/myStyle"
31          android:text="@string/minus"/>
32      <Button
33          style="@style/myStyle"
34          android:text="@string/arrow"/>
35      <Button
36          style="@style/myStyle"
37          android:text="@string/ce"/>
38      <Button
39          style="@style/myStyle"
40          android:text="@string/c"/>
41      <Button
42          style="@style/myStyle"
43          android:text="@string/plusminus"/>
44      <Button
45          style="@style/myStyle"
46          android:text="@string/correct"/>
47      <Button
48          style="@style/myStyle"
49          android:text="@string/seven"/>
50      <Button
51          style="@style/myStyle"
52          android:text="@string/eight"/>
53      <Button
54          style="@style/myStyle"
55          android:text="@string/nine"/>
56      <Button
57          style="@style/myStyle"
58          android:text="@string/div"/>
59      <Button
60          style="@style/myStyle"
61          android:text="@string/mode"/>
62      <Button
63          style="@style/myStyle"
64          android:text="@string/four"/>
65      <Button
66          style="@style/myStyle"
67          android:text="@string/five"/>
68      <Button
```

```
69          style="@style/myStyle"
70          android:text="@string/six"/>
71      <Button
72          style="@style/myStyle"
73          android:text="@string/mul"/>
74      <Button
75          style="@style/myStyle"
76          android:text="@string/daoshu"/>
77      <Button
78          style="@style/myStyle"
79          android:text="@string/one"/>
80      <Button
81          style="@style/myStyle"
82          android:text="@string/two"/>
83      <Button
84          style="@style/myStyle"
85          android:text="@string/three"/>
86      <Button
87          style="@style/myStyle"
88          android:text="@string/minus"/>
89      <Button
90          style="@style/myStyle"
91          android:text="@string/equal"
92          android:layout_rowSpan="2"
93          android:layout_height="100dp"/>
94      <Button
95          style="@style/myStyle"
96          android:text="@string/zero"
97          android:layout_columnSpan="2"
98          android:layout_width="120dp"/>
99      <Button
100         style="@style/myStyle"
101         android:text="@string/dot"/>
102     <Button
103         style="@style/myStyle"
104         android:text="@string/plus"/>
105 </GridLayout>
```

在字符串常量文件中定义了一些字符串常量，内容比较简单，请读者参考之前的代码，在此不再列出。

在样式文件 06\PageSwitcher\res\values\styles.xml 中添加如下代码

```
1 <style name="myStyle">
2     <item name="android:layout_width">60dp</item>
```

```
3        <item name="android:layout_height">50dp</item>
4        <item name="android:textSize">20sp</item>
5        <item name="android:gravity">center</item>
6    </style>
```

主程序代码：06\PageSwitcher\src\iet\jxufe\cn\android\pageswitcher\MainActivity.java

```
1   public class MainActivity extends Activity {
2       private ViewPager mViewPager;                  //存放多个页面的控件
3       private PagerTabStrip mTab;                    //页面选项卡
4       private LinearLayout mImgs;                    //存放底部图片的线性布局
5       private int[] layouts=new int[]{R.layout.linearlayout,
            R.layout.framelayout,R.layout.gridlayout };    //多个页面的布局文件
6       private String[] titles=new String[]{"按比例分割屏幕","霓虹灯",
            "计算器界面",};
7       private ImageView[] mImgViews=new ImageView[layouts.length];
8       private List<View>views=new ArrayList<View>();    //存放所有页面的集合
9       private List<String>pagerTitles=new ArrayList<String>();
                                                        //存放所有标题的集合
10      protected void onCreate(Bundle savedInstanceState){
11          super.onCreate(savedInstanceState);
12          getWindow().setFlags(WindowManager.LayoutParams.FLAG_FULLSCREEN,
            WindowManager.LayoutParams.FLAG_FULLSCREEN);   //设置屏幕为全屏显示
13          setContentView(R.layout.activity_main);
14          mImgs=(LinearLayout)findViewById(R.id.mImgs);   //根据ID获取相应的控件
15          mViewPager=(ViewPager)findViewById(R.id.mViewPager);
16          mTab=(PagerTabStrip)findViewById(R.id.mTab);
17          //设置选项卡之间的边距，默认情况下在一个页面中可以看见多个选项卡
            mTab.setTextSpacing(300);
18          init();                                     //执行初始化操作
19          mViewPager.setAdapter(new MyPagerAdapter());  //为ViewPager控件添
                                                        //加适配器
20          initImg();                                  //初始化显示的图片
21          //为ViewPager控件添加页面变换事件监听器
            mViewPager.setOnPageChangeListener(new MyPageChangeListener());
22      }
23      //自定义页面变换事件监听器
        private class MyPageChangeListener implements OnPageChangeListener {
24          public void onPageScrollStateChanged(int arg0){
25          }
26          public void onPageScrolled(int arg0,float arg1,int arg2){
27          }
28          public void onPageSelected(int selected){   //显示的页面发生变化时触发
29              resetImg();                             //重置底部显示的图片
```

```
30          //将当前页面对应的图片设置为红色
31          mImgViews[selected].setImageResource(R.drawable.choosed);
32        }
33     }
34     //自定义PagerAdapter类,该类用于封装需要切换的多个页面
       private class MyPagerAdapter extends PagerAdapter {
35        public int getCount(){                          //该方法返回所包含页面的个数
36          return views.size();
37        }
38        public boolean isViewFromObject(View arg0,Object arg1){
39          return arg0==arg1;
40        }
41        public CharSequence getPageTitle(int position){ //该方法用于返回页
                                                          //面的标题
42          return pagerTitles.get(position);
43        }
44        //该方法用于初始化指定的页面
          public Object instantiateItem(ViewGroup container,int position){
45          ((ViewPager)container).addView(views.get(position));
46          return views.get(position);
47        }
48        //该方法用于销毁指定的页面
          public void destroyItem(ViewGroup container,int position,Object object){
49          ((ViewPager)container).removeView(views.get(position));
50        }
51     }
52     public void init(){      //该方法用于初始化需要显示的页面,并将其添加到集合中
53        for(int i=0;i<layouts.length;i++){
54          View view=getLayoutInflater().inflate(layouts[i],null);
55          views.add(view);
56          pagerTitles.add(titles[i]);
57        }
58     }
59     public void initImg(){  //该方法用于初始化底部图片,并将图片添加到水平线性布局中
60        for(int i=0;i<mImgViews.length;i++){
61          mImgViews[i]=new ImageView(MainActivity.this);
62          if(i==0){                              //默认情况下第一张图被选中
63            mImgViews[i].setImageResource(R.drawable.choosed);
64          } else {
65            mImgViews[i].setImageResource(R.drawable.unchoosed);
66          }
67          mImgViews[i].setPadding(20,0,0,0);
68          mImgViews[i].setId(i);
69          mImgViews[i].setOnClickListener(mOnClickListener);
```

```
70              mImgs.addView(mImgViews[i]);
71          }
72      }
73      //声明一个匿名的单击事件处理器变量,用于处理底部图片的单击事件
        //主要是切换图片的显示,被单击的图片显示为红色,其他的显示为黄色
        private OnClickListener mOnClickListener=new OnClickListener(){
74          public void onClick(View v){
75              resetImg();              //重置图片的显示,将所有的图片都设置为黄色
76              //将被单击的图片设置为红色
                ((ImageView)v).setImageResource(R.drawable.choosed);
77              //切换页面的显示,根据单击的图片显示对应的页面
                mViewPager.setCurrentItem(v.getId());
78          }
79      };
80      public void resetImg(){//该方法用于将所有的ImageView都显示为未选中状态的图片
81          for(int i=0;i<mImgViews.length;i++){
82              mImgViews[i].setImageResource(R.drawable.unchoosed);
83          }
84      }
85  }
```

6.3 代码分析

6.3.1 界面分析

该程序界面中主要包含两个控件,即 ViewPager 和 LinearLayout,整体采用相对布局。其中,ViewPager 用于动态显示页面,它的宽度和高度占满整个屏幕,当滑动页面时,ViewPager 可以清除之前的页面,然后重新加载新的页面;LinearLayout 位于屏幕的下方,水平居中,用于显示图片控件,由于图片控件的内容是动态变化的,因此该部分信息需要在代码中动态确定。布局中 LinearLayout 的内容暂时为空,仅仅指定了线性布局的方向以及里面内容摆放的对齐方式,代码见 activity_main.xml。由于 ViewPager 类位于Android 提供的兼容包中,类似于第三方 JAR 包,不能直接使用,因此需要写完整的包名和类名,否则系统无法识别。

按比例分割屏幕、霓虹灯、计算器等页面引用了前面所介绍的一些例子的代码,在此只列出代码,不做详细说明,读者如有疑问,请查看前面案例中的详细说明。

6.3.2 ViewPager 介绍

ViewPager 使得用户可以很方便地在多个页面之间进行切换,只需要在页面中轻轻地向左或者向右滑动即可。ViewPager 本质上是一个容器,可以向该容器中添加控件或者删除控件。这里的控件可以是简单的控件,例如按钮、文本显示框等,也可以是包含复杂结构的控件,例如自定义的 View 对象、容器等。实际上,一个页面也可以看成是一个

复杂的 View 对象，通常将页面的设计放在布局文件中，之后调用 Activity 的 getLayout-Inflater().inflate()方法将布局文件转换成一个 View 对象，之后即可将该 View 对象添加到 ViewPager 容器中。

在 Android 中，ViewPager 对象本身并不直接与页面进行关联，而是通过一个中介或者适配器来关联，即 PagerAdapter。PagerAdapter 是一个抽象类，不能直接实例化，在 Android 系统中为用户提供了一些该类的实现类，例如 FragmentPagerAdapter、FragmentStatePagerAdapter。除此之外，用户也可以自定义该类的实现类，在此介绍最通用的方法，即自定义一个 PagerAdapter 的子类，该子类至少需要实现以下 4 个方法。

- instantiateItem(ViewGroup container, int position)：该方法用于初始化指定的页面。
- destroyItem(ViewGroup container, int position, Object object)：该方法用于销毁指定的页面。
- getCount()：该方法用于返回可切换的页面的个数。
- isViewFromObject(View arg0, Object arg1)：该方法用于判断 View 对象与 Object 对象是否为同一个对象。

除此之外，用户还可以重写 getPageTitle(int position)方法，用于获取指定页的标题等。自定义 PageAdapter 子类的代码见程序代码的第 34~51 行。

在创建完 PagerAdapter 类的对象之后，调用 ViewPager 对象的 setAdapter()方法即可将 ViewPager 对象与 PagerAdapter 对象关联起来，从而将页面与 ViewPager 结合起来。至此实现了页面的切换效果，但此时页面切换与下面的图片控件还没有建立对应关系。

我们希望页面切换后图片控件也发生变化，即程序能够监听到页面的变化，在此为 ViewPager 对象添加一个监听页面变化的事件监听器，即调用它的 setOnPageChangeListener()方法。该方法需要传递具体的监听器对象，即 OnPageChangeListener 接口的实现类，该监听器对象中包含 3 个方法，分别监听页面滚动时、页面滚动状态发生变化时以及页面被选中时的事件，在此只需要处理页面选中的事件即可。

在该事件处理中，首先让所有的图片控件都恢复为未选中状态对应的图标，然后将当前页面所对应的图片控件设置为选中的图标。具体代码见程序代码的第 23~33 行。

除了滑动可以切换页面之外，本程序还提供了单击下方的图片控件也可以切换到该图片所对应页面的功能。具体实现过程是初始化这些图片控件时为它们提供单击事件监听器，然后在单击事件处理中让页面切换到对应的页面。图片控件的初始化以及添加单击事件监听器的代码见第 59~72 行，单击事件的处理代码见第 73~84 行。

6.4 知识扩展

6.4.1 基于监听的事件处理

在本案例中涉及两种事件处理，即图片控件的单击事件处理以及 ViewPager 对象的页面发生变化的事件处理，这两种事件处理都是采用监听器进行监听，然后监听器中具体的方法进行处理。那么什么是监听器呢？基于监听的事件处理的流程又是怎样的呢？

Android 的基于监听的事件处理模型与 Java 的 AWT、Swing 的处理方式几乎完全一样,只是相应的事件监听器和事件处理方法名有所不同。在基于监听的事件处理模型中,主要涉及以下三类对象。

- **EventSource(事件源)**:即事件发生的源头,通常为某一控件,例如按钮、图片、列表等。
- **Event(事件)**:用户的具体某一操作的详细描述,事件封装了该操作的相关信息,如果程序需要获得事件源上所发生事件的相关信息,一般通过 Event 对象取得,例如按键事件按下的是哪个键、触摸事件发生的位置等。
- **EventListener(事件监听器)**:负责监听用户在事件源上的操作,并对用户的各种操作做出相应的响应。事件监听器中可包含多个事件处理器,一个事件处理器实际上就是一个事件处理方法。

那么在基于监听的事件处理中,这三类对象又是如何协作的呢?实际上,基于监听的事件处理是一种委托式事件处理,就是自己不做交给别人来做。普通控件(事件源)将整个事件处理委托给特定的对象(事件监听器),当该事件源发生指定的事件时,系统能够检测到并自动生成事件对象,然后通知所委托的事件监听器,由事件监听器根据事件类型调用相应的事件处理器来处理这个事件。事件处理模型如图 6-4 所示。

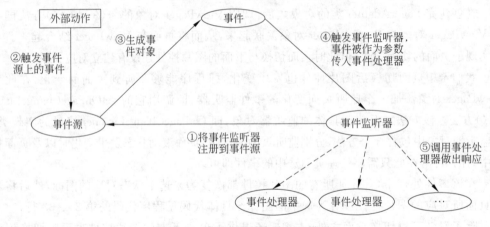

图 6-4 基于监听的事件处理模型

委托式事件处理非常类似于人类的社会分工,在生活中我们每个人的能力都有限,当**碰到一些自己处理不了的事情时就委托给某个机构或公司来处理**。此时,我们首先需要和该机构或公司建立联系,以方便及时沟通;其次,我们需要把遇到的事情或者要求向对方描述清楚,这样他们才能更好地设计解决方案,该机构可能比较大,不止处理我们的事情,因此会选派具体的员工来处理。其中,**我们自己就是事件源,遇到的具体事情就是事件,该机构就是事件监听器,具体解决事情的员工就是事件处理器。**

基于监听的事件处理模型的编程步骤如下:

(1) 获取事件源(即普通界面控件),也就是被监听的对象。

(2) 调用事件源的 set×××Listener()方法将事件监听器对象注册给事件源,即将事件源与事件监听器关联起来,这样当事件发生时就可以自动传递给事件监听器。

(3) 实现事件监听器类,该监听器类是一个特殊的 Java 类,必须实现一个×××Listener 接口,并实现接口里的所有方法,每个方法用于处理一种事件。

在上述步骤中,事件源比较容易获取,一般是界面控件,根据 findViewById()方法即可得到;调用事件源的 set×××Listener()方法是由系统定义好的,只需要传入一个具体的事件监听器,所以我们要做的就是实现事件监听器。所谓事件监听器,其实就是实现了特定接口的 Java 类的实例。在程序中实现事件监听器通常有以下几种形式。

- 内部类形式:将事件监听器类定义为当前类的内部类。
- 外部类形式:将事件监听器类定义成一个外部类。
- 类自身作为事件监听器类:让 Activity 本身实现监听器接口,并实现事件处理方法。
- 匿名内部类形式:使用匿名内部类创建事件监听器对象。

6.4.2 页面全屏显示

在默认情况下,手机界面由状态栏、标题栏以及页面内容 3 个部分组成,其中状态栏和标题栏是可以通过设置不显示的。在做 Android 应用,特别是游戏开发时,为了充分利用空间,也为了整个界面更加美观,往往会要求页面全屏显示,即不显示标题和状态栏。

(1) 在 Android 中不显示标题栏有两种方式,第一种是在 Java 代码中进行设置,在 setContentView()方法之前添加以下代码:

```
requestWindowFeature(Window.FEATURE_NO_TITLE);
```

第二种是在 AndroidManifest.xml 清单文件的 Activity 标签中设置 android:theme 属性,该属性值为@android:style/Theme.NoTitleBar,或者其他包含 NoTitleBar 的值。

(2) 在 Android 中不显示状态栏也有两种方式,第一种是在 Java 代码中进行设置,在 setContentView()方法之前添加以下代码:

```
getWindow().setFlags(WindowManager.LayoutParams.FLAG_FULLSCREEN,
    WindowManager.LayoutParams.FLAG_FULLSCREEN);
```

第二种是在 AndroidManifest.xml 清单文件的 Activity 标签中设置 android:theme 属性,该属性值为 @ android:style/Theme.NoTitleBar.Fullscreen,或者其他包含 Fullscreen 的值。

6.5 思考与练习

(1) 在本案例中只去除了状态栏,请尝试同时去除标题栏和状态栏。
(2) 在基于监听的事件处理模型中主要包含的三类对象是什么?
(3) 简单描述基于监听的事件处理的过程。
(4) 实现事件监听器的方式有_____、_____、_____和_____。
(5) Android 中基于监听的事件处理机制实现的基本思想应用了设计模式中的(   )。
  A) 观察者模式 B) 代理模式 C) 策略模式 D) 装饰者模式

第 7 章

图片定时滑动播放效果

7.1 案例概述

本案例主要实现图片定时滑动播放效果,每隔 8s,切换到另一张图片,同时可以看到图片滑动切换的过程,图片下方的圆圈对应着相应的图片,红色的圆圈对应当前的图片,图片变化时,圆圈的状态也会发生变化;同时,用户可直接单击某个圆圈,从而切换到相应的图片。除此之外,本案例也支持手势滑动操作,根据手势的方向与速度切换显示图片。本案例的程序运行效果如图 7-1 至图 7-3 所示。

图 7-1　程序运行效果图 1

图 7-2　程序运行效果图 2

图 7-3　程序运行效果图 3

7.2 关键代码

布局文件:07\ImageScan\res\layout\activity_main.xml

```
1   <LinearLayout xmlns:android="http://schemas.android.com/apk/res/android"
2       xmlns:tools="http://schemas.android.com/tools"
3       android:layout_width="match_parent"
4       android:layout_height="match_parent"
```

```
5         android:background="#aabbcc"
6         android:orientation="vertical">
7         <iet.jxufe.cn.android.imagescan.MyImageTopView
8             android:layout_width="match_parent"
9             android:layout_height="240dp"
10            android:id="@+id/mTopView">
11        </iet.jxufe.cn.android.imagescan.MyImageTopView>
12        <LinearLayout
13            android:layout_width="match_parent"
14            android:layout_height="wrap_content"
15            android:gravity="center"
16            android:orientation="horizontal"
17            android:id="@+id/mBottomView"/>
18    </LinearLayout>
```

程序代码:07\ImageScan\src\iet\jxufe\cn\android\imagescan\MainActivity.java

```
1   public class MainActivity extends Activity {
2       private MyImageTopView mTopView;        //自定义控件,用于显示上方图片的容器
3       private LinearLayout mBottomView;       //显示下方圆圈的容器(线性布局)
4       private int[] imgIds=new int[] {R.drawable.pic1,R.drawable.pic2,
            R.drawable.pic3,R.drawable.pic4,R.drawable.pic5,R.drawable.pic6,
            R.drawable.pic7};                   //需要显示的一组图片的ID
5       public ImageView[] imgViews=new ImageView[imgIds.length];
6       protected void onCreate(Bundle savedInstanceState){
7           super.onCreate(savedInstanceState);
8           setContentView(R.layout.activity_main);
9           mBottomView= (LinearLayout)findViewById(R.id.mBottomView);
10          mTopView= (MyImageTopView)findViewById(R.id.mTopView);
11          initBottom();                       //初始化底部的圆圈,默认第一个为选中
12          mTopView.initImages(imgIds);        //初始化要显示的图片
13      }
14      public void initBottom(){               //初始化底部的圆圈,并为圆圈添加单击事件处理
15          for(int i=0;i<imgViews.length;i++){
16              imgViews[i]=new ImageView(this);
17              if(i==0){
18                  imgViews[i].setImageResource(R.drawable.choosed);
19              } else {
20                  imgViews[i].setImageResource(R.drawable.unchoosed);
21              }
22              imgViews[i].setPadding(15,0,0,0);   //设置圆圈之间的边距
23              imgViews[i].setId(i);               //为每个圆圈添加ID
24              imgViews[i].setOnClickListener(new OnClickListener(){
25                  public void onClick(View v){
```

```
26              resetImg();                    //使所有的圆圈都不选中
27              ((ImageView)v).setImageResource(R.drawable.choosed);
                                               //将当前圆圈设为选中状态
28              mTopView.scrollToImage(v.getId());    //使图片切换到当前
                                                      //圆圈所对应的图片
29          }
30      });
31      mBottomView.addView(imgViews[i]); //将所有的圆圈都添加到线性布局中
32  }
33 }
34 public void resetImg(){         //该方法用于将所有的圆圈都显示为未选中状态
35      for(int i=0;i<imgViews.length;i++){
36          imgViews[i].setImageResource(R.drawable.unchoosed);
37      }
38  }
39 }
```

程序代码:07\ImageScan\src\iet\jxufe\cn\android\imagescan\MyImageTopView.java

```
1  public class MyImageTopView extends ViewGroup { //自定义控件,从 ViewGroup
                                                    //继承而来
2      private GestureDetector gesDetector;    //手势检测器
3      private Scroller scroller;              //滚动对象
4      private int currentImageIndex=0;        //记录当前显示的图片的序号
5      private boolean fling=false;            //添加标志,防止底层的 onTouch 事
                                               //件重复处理 UP 事件
6      private Handler mHandler;               //handler 对象,用于发送、接收和处理消息
7      private Context context;                //上下文对象
8      public MyImageTopView(Context context,AttributeSet attributeSet){
9          super(context,attributeSet);
10         this.context=context;
11         init();                             //执行初始化操作
12         this.setOnTouchListener(new MyOnTouchListener());//添加触摸事件处理
13     }
14     public void init(){
15         scroller=new Scroller(context);              //创建滚动条
16         mHandler=new Handler(){    //创建 Handler 对象,并重写其处理消息的方法
17             public void handleMessage(Message msg){
18                 if(msg.what==0x11){
19                     //收到消息后,切换到指定的图片
20                     scrollToImage((currentImageIndex+1)%getChildCount());
21                 }
22             }
23         };
```

```
24        gesDetector=new GestureDetector(context,new OnGestureListener(){
25            public boolean onSingleTapUp(MotionEvent e){
              //手指离开触摸屏的那一刹那调用该方法
26                return false;
27            }
28            public void onShowPress(MotionEvent e){
              //手指按在触摸屏上,它的时间范围在按下起效,在长按之前
29            }
30            public boolean onScroll(MotionEvent e1,MotionEvent e2,
31                    float distanceX,float distanceY){    //手指在触摸屏上滑动
32                //如果滑动范围在第一页和最后一页之间,distanceX>0 表示向右滑动
                  //distanceX<0 表示向左滑动,如果超出了这两个范围,则不做任何操作
                  if((distanceX>0 && getScrollX()<getWidth()
33                        * (getChildCount()-1))
34                        ||(distanceX<0 && getScrollX()>0)){
35                    scrollBy((int)distanceX,0);
                      //滚动的距离,在此只需要水平滚动,垂直方向滚动为 0
36                }
37                return true;
38            }
39            public void onLongPress(MotionEvent e){
              //手指按在屏幕上持续一段时间,并且没有松开
40            }
41            public boolean onFling(MotionEvent e1,MotionEvent e2,
42                    float velocityX,float velocityY){
                  //手指在触摸屏上迅速移动并松开
43                //判断是否达到最小滑动速度,取绝对值
44                if(Math.abs(velocityX)>ViewConfiguration.get(context)
45                        .getScaledMinimumFlingVelocity()){
                      //如果速度超过最小速度
46                    if(velocityX>0 && currentImageIndex>=0){
47                        fling=true;            //velocityX>0 表示向左滑动
48                        scrollToImage((currentImageIndex-1+getChildCount())
49                            %getChildCount());
50                    } else if(velocityX<0&& currentImageIndex<=
                          getChildCount()-1){
51                        fling=true;            //velocityX<0 表示向右滑动
52                        scrollToImage((currentImageIndex+1)%getChildCount());
53                    }
54                }
55                return true;
56            }
57            public boolean onDown(MotionEvent e){
              //手指刚刚接触到触摸屏的那一刹那,就是接触的那一下
```

```
58                return false;
59            }
60        });
61        Timer timer=new Timer();                        //创建定时器对象
62        timer.schedule(new TimerTask(){                 //定时操作,每隔8s发送消息
63            public void run(){
64                mHandler.sendEmptyMessage(0x11);
65            }
66        },0,8000);
67    }
68
69    public void scrollToImage(int targetIndex){         //跳转到目标图片
70        if(targetIndex !=currentImageIndex && getFocusedChild()!=null
71            && getFocusedChild()==getChildAt(currentImageIndex)){
72            getFocusedChild().clearFocus();             //当前图片清除焦点
73        }
74        final int delta=targetIndex * getWidth()-getScrollX(); //需要滑动的距离
75        int time =Math.abs(delta) * 5;
    //time 表示滑动的时间,单位为毫秒,滑动的时间是滑动距离的 5 倍
76        scroller.startScroll(getScrollX(),0,delta,0,time);
77        invalidate();                                   //刷新页面
78        currentImageIndex=targetIndex;                  //改变当前图片的索引
79        ((MainActivity)context).resetImg();
80        //改变下方圆圈的状态
81        ((MainActivity)context).imgViews[currentImageIndex].
            setImageResource(R.drawable.choosed);
82    }
83    public void computeScroll(){           //重写父类的方法,记录滚动条的新位置
84        super.computeScroll();
85        if(scroller.computeScrollOffset()){
86            scrollTo(scroller.getCurrX(),0);
87            postInvalidate();
88        }
89    }
90    private class MyOnTouchListener implements OnTouchListener{ //触摸事件监听器
91        public boolean onTouch(View v,MotionEvent event){
92            gesDetector.onTouchEvent(event);     //将触摸事件交由 GestureDetector
                                                   //处理
93            if(event.getAction()==MotionEvent.ACTION_UP){
94                if(!fling){                                //当用户停止拖动时
95                    snapToDestination();
96                }
97                fling=false;
98            }
```

```
 99                return true;
100            }
101        }
102        //该方法从ViewGroup中继承而来,是它的一个抽象方法,该方法用于指定容器里的
           //控件如何摆放,当控件大小发生变化时会回调该方法
           protected void onLayout(boolean changed,int left,int top,int right,
           int bottom){
103            for(int i=0;i<getChildCount();i++){    //设置布局,将子视图按顺序横向排列
104                View child=getChildAt(i);                //获取到每一个子控件
105                child.setVisibility(View.VISIBLE);    //设置该控件为可见
106                child.measure(right-left,bottom-top);
107                child.layout(i * getWidth(),0,(i+1) * getWidth(),getHeight());
108            }
109        }
110        private void snapToDestination(){            //滑动到指定图片
111            scrollToImage((getScrollX()+(getWidth()/ 2))/ getWidth());
               //四舍五入,若超过一半进入下一张图片
112        }
113        public void initImages(int[] imgIds){        //初始化显示的图片
114            int num=imgIds.length;                       //获取图片集合的长度
115            this.removeAllViews();                       //清空所有的控件
116            for(int i=0;i<num;i++){                      //循环逐个添加图片控件
117                ImageView imageView=new ImageView(getContext());
118                imageView.setImageResource(imgIds[i]); //设置每个图片控件的图片
119                this.addView(imageView);               //将图片添加到自定义的控件中
120            }
121        }
122    }
```

7.3 代码分析

7.3.1 界面分析

该界面中主要包含两个控件,即自定义的控件 MyImageTopView 和 LinearLayout,整体采用垂直的线性布局。其中,MyImageTopView 主要用于显示上方的可供切换的图片,它是自定义的一个控件,该控件从 ViewGroup 继承而来,是一个容器,可以存放多个控件,容器中控件的摆放是从左到右一个挨着一个摆放的;LinearLayout 位于 MyImageTopView 的下方,水平居中,用于显示圆圈,由于圆圈的状态是动态变化的,因此该部分信息需要在代码中动态确定。布局中 LinearLayout 的内容暂时为空,仅仅指定了线性布局的方向以及里面内容摆放的对齐方式,代码见 activity_main.xml。

由于 MyImageTopView 是用户自定义的控件,因此在布局文件中使用时必须用完整的包名+类名,否则系统无法识别。

7.3.2 自定义 MyImageTopView 控件

MyImageTopView 控件从 ViewGroup 类继承而来,需要实现其抽象方法 onLayout(),该方法用于指定容器里的控件如何摆放,当控件大小发生变化时会回调该方法。在此指定容器里的控件从左到右摆放,即控件的水平位置根据其在容器中的序号计算,而垂直位置都是从顶部开始。代码见 MyImageTopView.java 的第 102~109 行。

MyImageTopView 控件的主要作用是在多个图片之间滑动切换,首先需要为其初始化一些图片,在此为其单独写了一个方法 initImages(int[] imgIds),根据数组中的图片个数创建相应的图片控件,然后将其添加到容器中,代码见 MyImageTopView.java 的第 113~121 行。

定时滑动功能的实现主要是通过定时器来完成的,定时器每隔 8000ms 执行一次操作,主要是通过 Handler 对象发送一条消息,然后 Handler 对象接收消息,调用滑到下一张图片的操作。定时器的代码见第 61~66 行,消息处理的代码见第 16~23 行。

除了定时滑动之外,本案例也支持手势操作,因此需要为 MyImageTopView 控件添加触摸事件处理,然后在触摸事件处理方法中将触摸事件转交给手势检测器来处理。触摸事件处理代码见第 90~101 行,手势检测器的事件处理代码见第 24~60 行。手势检测主要识别滑动的距离、方向等,然后根据这个来计算具体滑到哪一张图片。除此之外,用户还需要重写父类的 computeScroll()方法及时将图片滑动到指定位置。

具体执行滑动操作的方法为 scrollToImage(int targetIndex),该方法需要传递一个参数,即目标图片的序号。其内部执行过程为首先计算出需要滑动的距离,然后设定滑动的时间,通常滑动的时间与滑动的距离有关,在此将滑动的时间设为滑动距离的 5 倍,最后调用滚动条的滑动方法 startScroll(),该方法需要传递 5 个参数,即起始点的 X 轴坐标、Y 轴坐标,X 轴滑动的距离、Y 轴滑动的距离,以及滑动的时间。滑动到目标页面后,底部的圆圈也要发生变化,而圆圈不属于 MyImageTopView 的一部分,因此涉及 MyImageTopView 与 MainActivity 之间的交互。在此,由于 MyImageTopView 创建时需要传递一个 Contxet 上下文参数,而该参数实际上就是 MainActivity,因此只需要将其进行强制类型转换,然后即可调用 MainActivity 的相关成员和方法。代码见第 69~82 行。

7.4 知识扩展

7.4.1 自定义控件

在 Android 中,所有的控件都是从 View 类继承而来的,在 View 类中提供了一些控件共同的属性,类似于 Java 中的 Object 类是所有类的超类一样。View 对象可以理解为屏幕上一块空白的矩形区域,不同的控件通过继承 View 类然后重写它的一些方法或者额外添加一些方法进行扩展,从而形成了风格迥异、功能强大的控件。基于这个原理,开发者完全可以通过继承 View 类或者 View 类的已有子类来创建具有自己风格的控件。

开发自定义控件的步骤如下。

(1) 自定义控件的类名,并让该类继承 View 类或一个现有的 View 类的子类。在本案例中希望创建一个可以容纳其他控件的容器控件,因此选择继承 View 类的子类 ViewGroup。

(2) 为自定义控件添加构造方法,然后选择性地重写父类的方法。在自定义控件时,需要提供显示的构造方法,构造方法是创建自定义控件的最基本的方式,无论是通过 Java 代码还是布局文件创建该控件,都会执行构造方法。而父类本身不存在无参数的构造方法,子类默认的无参数的构造方法会先调用父类的无参数的构造方法,从而会报错。除了构造方法必须提供以外,用户可以根据业务需要重写父类的部分方法。例如,onDraw()方法用于绘制界面。本案例中重写了父类的 onLayout()方法和 computeScroll()方法,这些方法是系统根据情况自动调用的。

(3) 在控件定义完之后,可以像系统控件一样使用它们,既可以在程序代码中通过该控件类创建控件对象,也可以在布局文件中通过标签创建控件。需要注意的是,在布局文件中控件的标签名是完整的包名+类名,而不仅仅是原来的类名,并且要求在定义该控件时在构造方法中要提供 AttributeSet 类型的参数。

7.4.2 手势检测

所谓手势,其实是指用户的手指或触摸笔在触摸屏上连续触碰的行为,例如在屏幕上从左到右滑这个动作。在 Android 中提供了手势检测,并为手势检测提供了相应的监听器。Android 中的手势检测类是 GestureDetector,在创建该类的对象时至少需要传递两个参数,即当前的上下文对象 Context 以及手势监听器 OnGestureListener。其中,手势监听器中包含若干个方法,分别用于监听用户的不同操作,主要有以下几个。

- boolean onDown(MotionEvent e):在手指刚刚接触到触摸屏的那一刹那触发该方法。
- void onShowPress(MotionEvent e):手指按在触摸屏上触发该方法,它的时间范围为从按下开始到长按之前。
- boolean onSingleTapUp(MotionEvent e):在手指离开触摸屏的那一刹那触发该方法。
- boolean onScroll(MotionEvent e1, MotionEvent e2, float distanceX, float distanceY):手指在触摸屏上滑动时触发该方法,该方法的 4 个参数分别表示滚动开始时的触摸事件、滚动结束时的触摸事件、X 轴滚动的距离、Y 轴滚动的距离。
- void onLongPress(MotionEvent e):手指按在屏幕上持续一段时间,并且没有松开时触发该方法。
- boolean onFling(MotionEvent e1, MotionEvent e2, float velocityX, float velocityY):手指在触摸屏上迅速移动,并松开时触发该方法,该方法 4 个参数分别表示开始移动时的触摸事件、松开时的触摸事件、X 轴的速度、Y 轴的速度。

使用 Android 中的手势检测只需要两个步骤:

（1）创建一个手势检测类 GestureDetector 的对象，并指定手势监听器。

（2）为当前页面或者特定控件添加触摸事件监听器，例如本案例中是为 MyImage-TopView 控件添加触摸事件监听器，然后在触摸事件处理中将触摸事件交给 GestureDetector 处理。

当用户触摸手机屏幕时，系统会自动生成一个触摸事件 MotionEvent，该事件将会被触摸事件监听器 OnTouchListener 监听到，然后会调用该监听器的 onTouch()方法，并将触摸事件作为参数传递给该方法，然而在 onTouch()方法内部又调用 GestureDetector 的 onTouch()方法，同样将触摸事件传递给它，一旦执行该方法，将会被手势监听器监听到，从而会调用手势监听器中的相关方法进行处理。

7.5 思考与练习

（1）请简单描述用户触摸屏幕到手势执行的过程。

（2）以下不是 OnGestureListener 接口中声明的方法的是（　　）。

  A）onDown() B）onUp() C）onScroll() D）onFling()

第8章 智能提示

8.1 案例概述

本案例设计和实现了一个简单的自动提示用户输入的效果。当用户输入一些字符后,系统会根据用户输入的内容自动匹配指定的数据源,并将所有符合要求的数据以列表的形式显示,用户可根据需要选择某一项完成输入。如果用户输入的内容在数据源中没有,则将用户输入的内容存入到数据源中,当再次输入时,会作为提示显示出来。本案例的程序运行效果如图 8-1 和图 8-2 所示。

图 8-1　程序运行效果图 1

图 8-2　程序运行效果图 2

8.2 关键代码

布局文件:08\AutoCompleteTest\res\layout\activity_main.xml

```
1   <LinearLayout xmlns:android="http://schemas.android.com/apk/res/android"
2       xmlns:tools="http://schemas.android.com/tools"
```

```
3       android:layout_width="match_parent"
4       android:layout_height="match_parent"
5       android:background="#aabbcc"
6       android:orientation="horizontal">
7       <AutoCompleteTextView
8           android:id="@+id/mAuto"
9           android:layout_width="0dp"
10          android:hint="@string/inputHint"
11          android:layout_height="wrap_content"
12          android:layout_weight="1"
13          android:completionThreshold="1"/>
14      <Button
15          android:layout_width="wrap_content"
16          android:layout_height="wrap_content"
17          android:onClick="search"
18          android:text="@string/search"/>
19  </LinearLayout>
```

字符串常量文件：08\AutoCompleteTest\res\values\strings.xml

```
1   <?xml version="1.0" encoding="utf-8"?>
2   <resources>
3       <string name="app_name">智能提示</string>
4       <string name="action_settings">Settings</string>
5       <string name="search">搜索</string>
6       <string name="inputHint">请输入搜索关键字</string>
7   </resources>
```

程序代码：08\AutoCompleteTest\src\iet\jxufe\cn\android\autocompletetest\MyOpenHelper.java

```
1   public class MyOpenHelper extends SQLiteOpenHelper {
2       public String createTableSQL="create table if not exists word" +
3               "(_id integer primary key autoincrement,word text)";
4       public MyOpenHelper(Context context,String name,CursorFactory
            factory,int version){
5           super(context,name,factory,version);
6       }
7       //在数据库创建后回调该方法,执行建表操作和插入初始化数据的操作
        public void onCreate(SQLiteDatabase db){
8           db.execSQL(createTableSQL);
9           db.execSQL("insert into word(word)values(?)",
                        new String[]{"Android 应用开发教程"});
10          db.execSQL("insert into word(word)values(?)",
                        new String[]{"疯狂 Android 讲义"});
```

```
11         db.execSQL("insert into word(word)values(?)",
                       new String[]{"Android开发揭秘"});
12         db.execSQL("insert into word(word)values(?)",
                       new String[]{"Android开发实战经典"});
13     }
14     //在数据库版本更新时回调该方法
       public void onUpgrade(SQLiteDatabase db,int oldVersion,int newVersion){
15         System.out.println("版本变化:"+oldVersion+"-------->"+newVersion);
16     }
17 }
```

程序代码：AutoCompleteTest\src\iet\jxufe\cn\android\autocompletetest\MainActivity.java

```
1  public class MainActivity extends Activity {
2      private List<String>datas;                  //用于存放数据源的集合
3      private AutoCompleteTextView mAuto;         //自动完成提示控件
4      private MyOpenHelper mHelper;               //数据库辅助类
5      private SQLiteDatabase mDB;                 //数据库封装类
6      private ArrayAdapter<String>adapter;        //关联数据与控件的适配器
7      protected void onCreate(Bundle savedInstanceState){
8          super.onCreate(savedInstanceState);
9          setContentView(R.layout.activity_main);
10         mHelper=new MyOpenHelper(this,"word.db", null, 1); //创建数据库辅助类
11         mDB=mHelper.getReadableDatabase();                 //获取数据库
12         mAuto= (AutoCompleteTextView)findViewById(R.id.mAuto);
13         datas=getData();                        //从数据库中获取数据
14         adapter=new ArrayAdapter<String>(this,android.R.layout.simple_
               list_item_1,datas);
15         mAuto.setAdapter(adapter);
16     }
17     public List<String>getData(){               //用于获取数据库中所有的数据源
18         List<String>contents =new ArrayList<String>();
19         Cursor result=mDB.rawQuery("select * from word",null);
20         while (result.moveToNext()){
21             contents.add(result.getString(result.getColumnIndex("word")));
22         }
23         return contents;
24     }
25     public void search(View view){              //"搜索"按钮的事件处理
26         String input=mAuto.getText().toString();
27         if(input==null||"".equals(input.trim())){
28             AlertDialog.Builder builder=new Builder(this);
29             builder.setTitle("警告提示");
30             builder.setMessage("请输入搜索关键字");
31             builder.create().show();
```

```
32          }else{
33              if(!datas.contains(input)){    //如果数据源中没有该关键字,则将该
                                               //关键字添加进去
34                  mDB.execSQL("insert into word(word)values(?)",
                        new String[]{input});
35                  datas=getData();
36                  adapter=new ArrayAdapter<String>(this,android.R.layout.
                        simple_list_item_1,datas);
37                  mAuto.setAdapter(adapter);
38              }
39          }
40      }
41      protected void onDestroy(){
42          if(mDB!=null){                              //退出时关闭数据库
43              mDB.close();
44          }
45          super.onDestroy();
46      }
47      public boolean onCreateOptionsMenu(Menu menu){
48          getMenuInflater().inflate(R.menu.main, menu);
49          return true;
50      }
51  }
```

8.3 代码分析

8.3.1 智能提示完成输入

智能提示完成输入是借助于 Android 中的 AutoCompleteTextView 控件完成的,该控件继承于 EditText,不同之处在于可根据用户输入的内容匹配指定的数据源,并将匹配的数据以列表的形式显示给用户,提示列表后,用户每输入一个字符就会从结果中再次进行筛选。AutoCompleteTextView 的常见属性如下。

- android:completionThreshold:设置最少输入的字符数,即用户至少输入几个字符后才会匹配数据源,显示提示信息,默认字符数为 2。
- android:completionHint:设置出现在信息列表中的提示信息。
- android:popupBackground:设置提示信息列表的背景。
- android:dropDownVerticalOffset:设置提示信息列表与文本框之间的垂直偏移像素,默认提示信息列表是紧跟着文本框的。
- android:dropDownHorizontalOffset:设置提示信息列表与文本框之间的水平偏移像素,默认提示信息列表与文本框左对齐。

有了控件之后,还需要为其指定数据源,数据源通常用数组或者集合表示,但实际数据往往存储在数据库或文件中,由于本案例中的数据源是动态变化的,所以选择使用集

合。集合中的数据又是从数据库中获取的,需要定义与数据库相关的操作类。在代码中 MyOpenHelper 类是一个自定义的数据库辅助类,主要用于创建数据库、更新数据库、获取数据库,该类从系统提供的数据库辅助类 SQLiteOpenHelper 继承而来,只需提供构造方法,重写其 onCreate()、onUpdate()方法即可。SQLiteDatabase 是数据库封装类,该类提供了数据库的一些常见操作,如增、删、查、改等,对于 SQLite 数据库的相关知识将在后面的案例中详细说明,在此不做介绍。

此时,数据源和控件是独立存在的,它们之间是如何关联的呢？也就是说,数据源是通过什么形式显示在控件之上的呢？在 Android 中为它们提供了一个中介——Adapter,在创建 Adapter 对象时,需要传递数据源、数据源中每一项显示的控件等参数,因此 Adapter 可以明确指定数据源的显示样式,然后为控件提供了一个 setAdapter()方法,将 Adapter 与控件关联起来。这样,数据源就和控件建立了关联,这种方式使得数据与显示数据的控件充分解耦,主要通过 Adapter 来控制数据的显示。

8.3.2 智能更新数据源

在用户输入内容后,单击"搜索"按钮,由系统判断数据源中是否存在该关键字,如果不存在,则添加到数据源中；如果存在,则不进行任何操作。在向数据库中插入数据时,调用的是 SQLiteDatabase 的相关方法,既可以通过 execSQL()方法传入插入记录的 SQL 语句,也可以通过 insert()方法传入表名、ContentValues 等参数。

当用户没有输入任何内容直接单击"搜索"按钮时,将弹出对话框,提示必须输入关键字,如图 8-3 所示。

图 8-3 弹出对话框

注意：在程序退出时应关闭数据库。

8.4 知识扩展

8.4.1 ArrayAdapter 介绍

ArrayAdapter 是系统提供的 BaseAdapter 抽象类的一个子类,通常用于存放数组或集合元素。在默认情况下,将元素内容显示在一个 TextView 上,如果数组或集合的元素不是字符串而是对象,则会调用对象的 toString()方法将其结果显示在 TextView 上。如果想将结果显示在其他 View 控件上,例如 ImageView 上,则需要重写 ArrayAdapter 对象的 getView()方法,将其返回类型设置为 ImageView。创建 ArrayAdapter 对象通常需要传递下面 3 个参数。

(1) 第一个参数为 Context 对象,即当前控件所在的上下文对象,通常是当前的 Activity。

(2) 第二个参数用于指定文本内容的显示样式,是一个 TextView 控件,可以使用系统提供的也可以自定义。本案例中的 android. R. layout. simple_list_item_1 就是调用系统中的一个布局文件，该文件只有一个标签，即＜TextView＞，如果想用自定义的 TextView，只需创建一个布局文件，该布局文件中只有一个标签就是＜TextView＞，然后在该标签内设置各种属性，例如大小、颜色等。

（3）第三个参数指定数据的来源，既可以是数组，也可以是集合。

在创建 ArrayAdapter 对象时，除了传入 3 个参数外，还可以传 4 个参数，即当布局文件中包含多个标签时，不仅需要指定布局文件，还需要指定 TextView 控件的 ID，即 ArrayAdapter(上下文对象，布局文件，TextView 控件 ID，数据源)。

8.4.2 对话框

对话框是一种漂浮在 Activity 之上的小窗口，在弹出对话框时，Activity 会失去焦点，对话框获取用户的所有交互。对话框常用于通知，它会临时打断用户，执行一些与应用程序相关的小任务，如任务的执行进度或登录提示等。

AlertDialog 是 Dialog 的子类，它能创建大部分用户交互的对话框，也是系统推荐的对话框类型。通常使用 AlertDialog 创建对话框，大致步骤如下：

（1）创建 AlertDialog.Builder 对象，该对象是 AlertDialog 的创建器。
（2）调用 AlertDialog.Builder 的方法，为对话框设置图标、标题、内容等。
（3）调用 AlertDialog.Builder 的 create()方法，创建 AlertDialog 对话框。
（4）调用 AlertDialog 的 show()方法，显示对话框。

在上述步骤中，主要是 AlertDialog 的内部类 Builder 在起作用，下面我们来看看 Builder 类中有哪些方法。Builder 内部类的主要方法如表 8-1 所示。

表 8-1 Builderl 类的主要方法表

方法签名	作用
public Builder setTitle	设置对话框标题
public Builder setMessage	设置对话框内容
public Builder setIcon	设置对话框图标
public Builder setPositiveButton	添加肯定按钮（Yes）
public Builder setNegativeButton	添加否定按钮（No）
public Builder setNeutralButton	添加普通按钮
public Builder setOnCancelListener	添加取消监听器
public Builder setCancelable	设置对话框是否可取消
public Builder setItems	添加列表
public Builder setMultiChoiceItems	添加多选列表
public Builder setSingleChoiceItems	添加单选列表
public AlertDialog create()	创建对话框
public AlertDialog show()	显示对话框

注意：表中大部分方法的返回值的类型都是 Builder 类型，也就是说，调用 Builder 对象的这些方法后返回的是该对象本身。用户可以把初始的 Builder 对象理解为一个空壳子，调用 Builder 对象的一个方法就是往这个壳子里面的指定位置添加一些东西，由于每一样东西的位置都是固定的，因此，如果多次调用同一个方法，则后面的值会覆盖前面的值。调用 Builder 对象方法的过程就是构造对话框的过程，一旦构建完成，即可创建、显示出来。

8.5 思考与练习

（1）本案例中的列表项使用的是系统提供的样式 android.R.layout.simple_list_item_1,请自定义一个样式，使弹出的列表项的文字颜色为红色，大小为 24sp，项与项之间的距离为 20dp。

（2）AutoCompleteTextView（自动完成输入）控件可根据用户输入的内容从指定的数据源中匹配出所有符合条件的数据，并以列表的形式显示，从而让用户进行选择，通过以下（　　）属性可以设置弹出列表所需要用户输入的最少字符数。

  A）android:completionThreshold B）android:completionHint

  C）android:dropDownVerticalOffset D）android:dropDownHorizontalOffset

（3）在下列选项中，前后两个类不存在继承关系的是（　　）。

  A）TextView、AutoCompleteTextView

  B）TextView、Button

  C）ImageView、ImageSwitcher

  D）ImageView、ImageButton

（4）Android 中包含了很多 Adapter 的相关类，在下列选项中，不是从 BaseAdapter 继承而来的类是（　　）。

  A）ArrayAdapter B）SimpleAdapter

  C）CursorAdapter D）PagerAdapter

（5）下列关于 AlertDialog 的描述不正确的是（　　）。

  A）AlertDialog 的 show()方法可创建并显示对话框

  B）AlertDialog.Builder 的 create() 和 show()方法都返回 AlertDialog 对象

  C）AlertDialog 不能直接用 new 关键字构建对象，而必须使用其内部类 Builder

  D）AlertDialog.Builder 的 show()方法可创建并显示对话框

（6）在构建 AlertDialog 时需要借助其内部类 Builder，Builder 类中包含了很多方法，在下列方法中，方法的返回值类型与其他项不同的是（　　）。

  A）create() B）setMessage()

  C）setView() D）setAdapter()

（7）AlertDialog 对话框中按钮的个数最多可以有（　　）个。

  A）1 B）2 C）3 D）无数

第 9 章

仿画廊视图效果

9.1 案例概述

本案例主要介绍综合使用水平滚动条和水平线性布局实现画廊效果。Android 前期版本中提供了 Gallery 控件,该控件可以很方便地实现图片浏览功能,遗憾的是该类在 Android 4.1 中被废弃了,官方文档推荐使用水平滚动条或者 ViewPager 来实现该效果。本案例通过水平滚动条模仿画廊效果,在选中底部列表中的某一图片后,上方的图片控件将会显示该图,底部列表中被选中的图片完全显示,未选中的图片以半透明的形式显示。本案例的程序运行效果如图 9-1 和图 9-2 所示。

图 9-1 程序运行效果图 1

图 9-2 程序运行效果图 2

9.2 关键代码

布局文件:09\ScrollViewGallery\res\layout\activity_main.xml

```
1    <LinearLayout xmlns:android="http://schemas.android.com/apk/res/android"
2        xmlns:tools="http://schemas.android.com/tools"
```

```
3      android:layout_width="match_parent"
4      android:layout_height="match_parent"
5      android:background="#aabbcc"
6      android:orientation="vertical" >
7      <ImageSwitcher
8          android:id="@+id/mSwitcher"
9          android:layout_width="wrap_content"
10         android:layout_height="0dp"
11         android:layout_weight="1"
12         android:paddingBottom="10dp"
13         android:inAnimation="@android:anim/fade_in"
14         android:outAnimation="@android:anim/fade_out"/>
15     <HorizontalScrollView
16         android:layout_width="wrap_content"
17         android:layout_height="wrap_content">
18         <LinearLayout
19             android:id="@+id/mLinear"
20             android:orientation="horizontal"
21             android:layout_width="wrap_content"
22             android:layout_height="wrap_content">
23         </LinearLayout>
24     </HorizontalScrollView>
25 </LinearLayout>
```

自定义边框图形文件：09\ScrollViewGallery\res\drawable-hdpi\bg.xml

```
1  <?xml version="1.0" encoding="utf-8"?>
2  <shape xmlns:android="http://schemas.android.com/apk/res/android" >
3      <solid android:color="#00000000"/>
4      <stroke android:color="#ff0000"
5          android:width="2dp"/>
6  </shape>
```

程序代码：09\ScrollViewGallery\src\iet\jxufe\cn\android\scrollviewgallery\MainActivity.java

```
1  public class MainActivity extends Activity {
2      private LinearLayout mLinear;              //存放底部缩略图的线性布局
3      private ImageSwitcher mSwitcher;           //图片切换器
4      List<Integer> imgIds;                      //存放所有图片ID的集合
5      private ImageView[] imgViews;              //显示图片的控件
6      protected void onCreate(Bundle savedInstanceState){
7          super.onCreate(savedInstanceState);
8          setContentView(R.layout.activity_main);
9          mLinear=(LinearLayout)findViewById(R.id.mLinear);
```

```
10      mSwitcher=(ImageSwitcher)findViewById(R.id.mSwitcher);
11      mSwitcher.setFactory(new ViewFactory(){  //图片切换器的初始化
12          public View makeView(){
13              ImageView img=new ImageView(MainActivity.this);
14              return img;
15          }
16      });
17      imgIds=getImageIds();                    //获取所有需要显示的图片
18      mSwitcher.setImageResource(imgIds.get(0));   //默认显示第一张图片
19      init();                                  //执行初始化操作
20  }
21  public void init(){
22      imgViews=new ImageView[imgIds.size()];   //创建一个存放图片控件的数组
23      //创建一个布局参数,指定控件的大小和边距
        LinearLayout.LayoutParams layoutParams = new LinearLayout.LayoutParams(60,80);
24      layoutParams.setMargins(0, 0, 5, 0);     //设置图片之间的边距
25      //为每张图片创建一个ImageView控件,并对其进行简单设置
        for(int i=0; i<imgViews.length; i++){
26          imgViews[i]=new ImageView(this);     //创建ImageView控件
27          imgViews[i].setId(imgIds.get(i));    //为ImageView控件添加ID属性
28          imgViews[i].setBackgroundResource(R.drawable.bg);
            //为ImageView控件设置背景边框
29          imgViews[i].setImageResource(imgIds.get(i));
            //为ImageView控件设置图片
30          imgViews[i].setLayoutParams(layoutParams);
31          imgViews[i].setOnClickListener(new MyListener());
32          if(i!=0){                            //默认第一张图片完全显示,其他图片半透明显示
33              imgViews[i].setImageAlpha(100);
34          }else{
35              imgViews[i].setImageAlpha(255);
36          }
37          mLinear.addView(imgViews[i]);        //将ImageView控件添加到线性布局中
38      }
39  }
40  private class MyListener implements OnClickListener{  //底部图片单击事件监听器
41      public void onClick(View v){
42          mSwitcher.setImageResource(v.getId());
43          setAlpha(imgViews);
44          ((ImageView)v).setImageAlpha(255);
45      }
46  }
47  public void setAlpha(ImageView[] imageViews){  //设置所有图片的Alpha值为100
48      for(int i=0; i<imageViews.length; i++){
```

```
49              imageViews[i].setImageAlpha(100);
50          }
51      }
52      public List<Integer>getImageIds(){      //通过反射机制获取所有符合条件的图片 ID
53          List<Integer>imageIds=new ArrayList<Integer>();
54          try {                               //获取 R.drawable 类中的所有成员变量
55              Field[] drawableFields=R.drawable.class.getFields();
                //循环遍历这些成员变量,然后判断是否符合指定条件
                //如果符合条件则将其值添加到集合中
56
57              for(Field field : drawableFields){
58                  if(field.getName().startsWith("x_")){
59                      imageIds.add(field.getInt(R.drawable.class));
60                  }
61              }
62          } catch(Exception e){
63              e.printStackTrace();
64          }
65          return imageIds;
66      }
67  }
```

9.3 代码分析

9.3.1 界面分析

该界面中主要包含两个部分,上方为用于显示大图的 ImageSwitcher,下方为用于显示所有图片缩略图的水平线性布局。由于图片较多,屏幕的宽度无法容纳所有的图片,在默认情况下,水平线性布局中超出屏幕边界的控件将无法显示。为了能让所有的图片都可以在屏幕上显示,这里在水平线性布局外面添加一个水平滚动条,然后将水平线性布局放入该滚动条内部。通过横向拖动滚动条,即可查看超出屏幕部分的图片。

由于图片控件的多少需要根据图片的个数来确定,图片控件的状态也是根据用户操作来动态变化的,所以不宜在布局文件中添加和指定。因此,布局中 LinearLayout 的内容暂时为空,仅仅指定了线性布局的方向,代码见 activity_main.xml。

9.3.2 ImageSwitcher 介绍

ImageSwitcher(图片切换器)主要用于图片间的切换,可以显示图片,与 ImageView 的不同之处在于切换显示图片时可以为图片添加进入时和退出时动画。

既然是切换那么肯定是在两个视图之间进行的,ImageSwitcher 通过 setFactory()方法来创建两个需要切换的视图。该方法需要传递一个 ViewFactory 类型的参数,该参数是一个工厂接口,专门用于创建控件,该接口内部只有一个方法——makeView(),用于返回所创建的控件。在实现 ViewFactory 接口时,必须要实现 makeView()方法,作为图片

切换器，makeView（）方法应该返回能够显示图片的控件，在此为 ImageView。在 setFactory（）方法内部实际上调用了两次 ViewFactory 接口的 makeView（）方法，从而创建了两个 ImageView 控件进行图片的切换。

因此，在创建 ImageSwitcher 对象之后，还必须调用其 setFactory（）方法对它进行初始化，否则无法实现切换功能。在图片切换时可以为其添加进入时和退出时动画，有两种方式。

（1）在布局文件中，为 ImageSwitcher 标签添加 android：inAnimation 和 android：outAnimation 属性分别设置进入时的动画和退出时的动画，既可以引用系统提供的动画，也可以是用户自定义的动画。例如，淡入淡出效果的设置如下：

```
android:inAnimation="@android:anim/fade_in"
android:outAnimation="@android:anim/fade_out"
```

（2）在 Java 代码中，调用 ImageSwitcher 对象的 setInAnimation（）和 setOutAnimation（）方法，将相应的动画传递进去即可。例如：

```
mSwitcher.setInAnimation (AnimationUtils.loadAnimation (this, android.R.anim.fade_in));
mSwitcher.setOutAnimation (AnimationUtils.loadAnimation (this, android.R.anim.fade_out));
```

当需要切换到下一张图片时，只需要调用 mSwitcher.setImageResource（）传入一个图片资源 ID，该图片资源就是即将显示的图片。

9.4 知识扩展

在前面图片切换的案例中，我们将所有的图片 ID 保存在一个数组中，在对这个数组进行初始化时，需要将每个图片的 ID 都添加进去，如果图片较多，代码比较冗长，特别是在新添加图片或对图片的名称进行修改时，还需要对程序的源代码进行修改，扩展性和灵活性不太好。实际上，可以通过 Java 的反射机制动态获取 Android 中的图片 ID。这是因为 Android 中的图片文件都会在 R.drawable 类中生成资源 ID，并且一个图片的文件对应于 R.drawable 类的一个成员变量，只需要获取 R.drawable 类中的所有成员变量，即可获取所有图片的 ID，然后对成员变量名进行判断即可获取符合要求的图片。而对于 Java 反射机制而言，获取成员变量名非常简单。

所谓的反射是指在运行状态中对于任意一个类都能够知道这个类的所有属性和方法，对于任意一个对象都能够调用它的任意一个方法和属性。Java 反射机制的关键就是要得到用户想要探索的类的 Class 对象，有了 Class 对象之后就可以进一步获取该类的成员变量、方法、构造方法等，在 Java 中与反射相关的 API 存放在 java.lang.reflect 包下。在 Java 程序中获取 Class 对象通常有以下 3 种方式。

（1）使用 Class 类的 forName（）静态方法：该方法需要传递一个字符串参数，该字符串为某个类的完整包名＋类名。

(2) 调用某个类的 class 属性来获取该类对应的 Class 对象：本案例中使用 R.drawable.class 将会返回 R.drawable 类对应的 Class 对象。

(3) 调用某个对象的 getClass() 方法：该方法是 java.lang.Object 类中的一个方法，所有的类都可以调用该方法，该方法将返回该对象所属类对应的 Class 对象。

在获取 R.drawable 类对应的 Class 对象之后，即可调用它的 getFields() 方法获取 R.drawable 类中所有的公共成员变量，以及调用 Field 对象的 getName() 方法获取成员变量名，即图片的文件名，然后判断是否为需要的图片，如果是，则获取该成员变量的值，即图片的 ID。

9.5 思考与练习

(1) 请修改现有程序，在 ImageSwitcher 控件的左、右添加两个按钮，单击按钮能够浏览上一张和下一张图片，同时下方的缩略图也会随之发生变化。

(2) 以下不是存放在 java.lang.reflect 包下的类是（   ）。
　　A) Class　　　　B) Field　　　　C) Method　　　　D) Constructor

(3) 假设定义了一个 LinearLayout 线性布局，在布局中添加控件的方法是（   ）。
　　A) addAction()　　B) addView()　　C) addChild()　　D) addLayout()

第10章 南昌景点介绍

10.1 案例概述

本案例主要介绍 ListView 列表控件的使用，当应用中包含多项数据，并且每项数据结构相同，只是内容不同时，通常需要通过列表来显示，列表中的内容可以是很简单的显示字符串的 TextView，也可以是结构比较复杂的包含多个控件的容器。本案例中通过列表显示南昌常见景点的基本信息，单击列表中的某一项后可以查看该景点的详细信息，图片徐徐展开。本案例涉及复杂列表项的构建、列表项的事件处理、页面的跳转、数据的传递等。本案例的程序运行效果如图 10-1 和图 10-2 所示。

图 10-1 程序运行效果图 1

图 10-2 程序运行效果图 2

10.2 关键代码

主界面布局文件：10\SceneryInfo\res\layout\activity_main.xml

```
1  <LinearLayout xmlns:android="http://schemas.android.com/apk/res/android"
2      xmlns:tools="http://schemas.android.com/tools"
```

```
3       android:layout_width="match_parent"
4       android:layout_height="match_parent"
5       android:background="#aabbcc"
6       android:orientation="vertical">
7       <TextView
8           android:layout_width="match_parent"
9           android:layout_height="wrap_content"
10          android:text="@string/title"
11          android:textSize="24sp"
12          android:background="#ccbbaa"
13          android:textColor="#000000"
14          android:padding="10dp"
15          android:gravity="center" />
16      <ListView
17          android:id="@+id/scenery"
18          android:layout_width="match_parent"
19          android:layout_height="wrap_content"
20          android:divider="#aaaaaa"
21          android:dividerHeight="2dp"
22          android:gravity="center" />
23  </LinearLayout>
```

列表中每一项的布局文件：10\SceneryInfo\res\layout\item.xml

```
1   <LinearLayout xmlns:android="http://schemas.android.com/apk/res/android"
2       xmlns:tools="http://schemas.android.com/tools"
3       android:layout_width="match_parent"
4       android:layout_height="match_parent"
5       android:layout_margin="10dp"
6       android:orientation="horizontal" >
7       <ImageView
8           android:id="@+id/image"
9           android:layout_width="100dp"
10          android:layout_height="75dp"
11          android:contentDescription="@string/imgInfo" />
12      <LinearLayout
13          android:layout_width="wrap_content"
14          android:layout_height="wrap_content"
15          android:layout_margin="10dp"
16          android:orientation="vertical" >
17          <TextView
18              android:id="@+id/name"
19              android:layout_width="match_parent"
20              android:layout_height="wrap_content"
```

```
21              android:textColor="#000000"
22              android:textSize="20sp" />
23          <TextView
24              android:id="@+id/brief"
25              android:layout_width="wrap_content"
26              android:layout_height="wrap_content"
27              android:layout_marginTop="10dp"
28              android:gravity="left"
29              android:singleLine="true"
30              android:ellipsize="end"
31              android:textColor="#0000ee"
32              android:textSize="12sp" />
33      </LinearLayout>
34  </LinearLayout>
```

显示景点详细信息的布局文件:10\SceneryInfo\res\layout\scenery_show.xml

```
1   <ScrollView xmlns:android="http://schemas.android.com/apk/res/android"
2       android:layout_width="match_parent"
3       android:layout_height="match_parent" >
4       <LinearLayout
5           android:layout_width="match_parent"
6           android:layout_height="wrap_content"
7           android:background="#aabbcc"
8           android:orientation="vertical" >
9           <TextView
10              android:id="@+id/title"
11              android:layout_width="match_parent"
12              android:layout_height="wrap_content"
13              android:gravity="center"
14              android:padding="10dp"
15              android:background="#ccbbaa"
16              android:textSize="24sp"/>
17          <ImageView
18              android:id="@+id/image"
19              android:layout_width="320dp"
20              android:layout_height="240dp"
21              android:scaleType="fitXY"
22              android:contentDescription="@string/imgInfo"/>
23          <TextView
24              android:id="@+id/content"
25              android:layout_width="match_parent"
26              android:layout_height="wrap_content"
27              android:background="#333333"
```

```
28            android:textColor="#00ff00"
29            android:textSize="16sp"
30            android:padding="10dp"/>
31     </LinearLayout>
32  </ScrollView>
```

程序代码：10\SceneryInfo\src\iet\jxufe\cn\android\sceneryinfo\MainActivity.java

```
1   public class MainActivity extends Activity {
2       private ListView mScenery;                    //显示景点信息的列表
3       //保存每个景点信息的集合
        private List<Map<String,Object>>list=new ArrayList<Map<String,
            Object>>();
4       private int[] imgIds=new int[] { R.drawable.tengwangge,
            R.drawable.badashanren,R.drawable.hanwangfeng,
            R.drawable.xiangshangongyuan,R.drawable.xishanwanshougong,
            R.drawable.meiling };              //存放景点图片ID的数组
5       private String[] names=new String[] { "滕王阁","八大山人纪念馆","罕王
            峰","象山森林公园","西山万寿宫","梅岭" };  //存放景点名称的数组
6       private String[] briefs=new String[] { "江南三大名楼之首","集收藏、陈列、
            研究、宣传为一体","青山绿水,风景多彩,盛夏气候凉爽","避暑、休闲、疗养、度假的最
            佳场所","江南著名道教宫观和游览胜地","山势嵯峨,层峦叠翠,四时秀色,气候宜人"
            };                                  //存放景点简单介绍信息的数组
7       private int[] contentIds=new int[]{R.raw.tengwg_info,R.raw.badsr_info,
            R.raw.hangwf_info,R.raw.xiangsgy_info,R.raw.wansg_info,
            R.raw.meil_info };                 //存放景点详细介绍文件ID的数组
8       protected void onCreate(Bundle savedInstanceState){
9           super.onCreate(savedInstanceState);
10          setContentView(R.layout.activity_main);
11          mScenery= (ListView)findViewById(R.id.scenery);
12          init();        //执行初始化操作,将同一个景点的图片、名称、简介、详情关联起来
13          //创建Adapter对象,将数据源与显示的控件关联起来
            SimpleAdapter adapter=new SimpleAdapter(this,list,R.layout.item,
                new String[] { "img","name","brief" },new int[] {
                R.id.image,R.id.name,R.id.brief });
14          mScenery.setAdapter(adapter);
15          //为列表添加列表项单击事件监听器
            mScenery.setOnItemClickListener(new OnItemClickListener(){
16          //列表项单击事件处理,跳转页面、传递数据
            public void onItemClick(AdapterView<?>parent,View view,int
                position,long id){
17              Intent intent=new Intent();
18              intent.setClass(MainActivity.this,SceneryShowActivity.class);
19              intent.putExtra("name",(String)list.get(position).get("name"));
```

```
20              intent.putExtra("image",(Integer)list.get(position).get("img"));
21              intent.putExtra("content",(Integer)list.get(position).
                    get("content"));
22              startActivity(intent);        //跳转页面
23          }
24      });
25  }
26  private void init(){              //将每个景点的图片、名称、简介、详情关联起来
27      for(int i=0;i<imgIds.length;i++){
28          Map<String,Object>item=new HashMap<String,Object>();
29          item.put("img",imgIds[i]);           //添加景点图片ID
30          item.put("name",names[i]);           //添加景点名称
31          item.put("brief",briefs[i]);         //添加景点简介
32          item.put("content",contentIds[i]);   //添加景点内容
33          list.add(item);                      //添加景点
34      }
35  }
36 }
```

程序代码：10\SceneryInfo\src\iet\jxufe\cn\android\sceneryinfo\SceneryShowActivity.java

```
1  public class SceneryShowActivity extends Activity {
2      private TextView title,content;          //显示标题和内容的TextView
3      private ImageView imageView;             //显示图片的ImageView
4      private ClipDrawable clipDrawable;       //用于图片徐徐展开
5      private Handler mHandler;
6      protected void onCreate(Bundle savedInstanceState) {
7          super.onCreate(savedInstanceState);
8          setContentView(R.layout.scenery_show);
9          imageView=(ImageView)findViewById(R.id.image);
10         content=(TextView)findViewById(R.id.content);
11         title=(TextView)findViewById(R.id.title);
12         //获取传递过来的数据,包括标题、图片ID、详情文件ID
           String titleStr=getIntent().getStringExtra("name");
13         int image=getIntent().getIntExtra("image",R.drawable.tengwangge);
14         int info=getIntent().getIntExtra("content",R.raw.tengwg_info);
15         title.setText(titleStr);
16         //创建ClipDrawable对象,指定裁剪的图片、方向等
           clipDrawable =new ClipDrawable(getResources().getDrawable(
               image),Gravity.CENTER,ClipDrawable.HORIZONTAL);
17         imageView.setImageDrawable(clipDrawable);
18         startThread();                       //启动线程,开始裁剪
19         //根据文件ID获取输入流
           InputStream inputStream=getResources().openRawResource(info);
```

```
20        content.setText(getStringFromInputStream(inputStream));
21        mHandler=new Handler(){                    //创建 Handler 对象
22            public void handleMessage(Message msg) {   //处理消息
23                clipDrawable.setLevel(clipDrawable.getLevel()+400);
                  //让图片可显示部分不断增大
24            }
25        };
26    }
27    public String getStringFromInputStream(InputStream inputStream){
      //从输入流中读取字符串
28        byte[] buffer=new byte[1024];
29        int hasRead=0;                             //记录读取的字节个数
30        StringBuilder result=new StringBuilder("");
31        try{
32            while((hasRead=inputStream.read(buffer))!=-1){
33                //根据读取的字节构建字符串,并添加到已有的字符串后面
                  result.append(new String(buffer, 0,hasRead,"GBK"));
34            }
35        }catch (Exception ex){
36            ex.printStackTrace();
37        }
38        return result.toString();
39    }
40    public void startThread(){
41        clipDrawable.setLevel(0);                  //设置图片初始不可见
42        new Thread(){                              //创建线程
43            public void run() {
44                while(clipDrawable.getLevel()<10000){   //判断图片是否完全显示
45                    try {
46                        Thread.sleep(300);                 //休眠 0.5s
47                        mHandler.sendEmptyMessage(0x11);   //发送空消息
48                    } catch (Exception e) {
49                        e.printStackTrace();
50                    }
51                }
52            }
53        }.start();                                 //启动线程
54    }
55 }
```

10.3 代码分析

10.3.1 界面分析

本案例中包含两个界面,一个是用于显示南昌各景点基本信息的主界面,另一个是用

于显示单个景点详细信息的界面,单击主界面中的某一项后可以跳转到对应的详细信息界面。

在主页面中主要包含两个控件,即显示标题的 TextView 以及显示景点信息的 ListView,整体采用垂直线性布局。在 ListView 中,设置了项与项之间分割线的颜色以及分割线的大小,即 android:divider 和 android:dividerHeight 属性,代码见 activity_main.xml 文件。列表中的每一项又包含 3 项信息,即景点图片、景点名、景点简介,这 3 项信息整体放在一个水平的线性布局中,其中又嵌套了一个垂直的线性布局,效果如图 10-3 所示,代码见 item.xml 文件。

图 10-3 列表项分析图

在详细信息界面中主要包含 3 个控件,即显示标题的 TextView、显示图片的 ImageView、显示景点信息的 TextView,整体采用垂直线性布局,代码见 scenery_show.xml 文件。

注意:创建一个新的 Activity 之后,在使用之前一定要在清单文件中对其进行注册。

10.3.2 ListView 介绍

ListView 是 Android 中使用非常广泛的一种控件,它以垂直列表的形式显示所有的列表项。在 Android 中使用 ListView 控件与使用其他控件类似,通常是在布局文件中添加 ListView 标签,然后在代码中根据 ID 获取 ListView 控件,之后即可为其设置数据源、显示数据等。除此之外,在 Android 中还专门提供了 ListActivity,其内部包含了一个 ListView,因此只需要让当前的 Activity 从 ListActivity 继承即可获得一个 ListView,而不需要布局文件。

在获取 ListView 控件之后,关键是为其指定数据源,但 ListView 本身并不能直接与需要显示的数据源关联,需要借助中介 Adapter 的协助,然后通过 setAdapter()方法将 Adapter 与列表关联起来。Adapter 主要用于关联数据源与所显示控件之间的关系。通过查看 Android 的帮助文档,总结出 Adapter 层次结构如图 10-4 所示。

Adapter 层次结构中比较关键的类就是 BaseAdapter 类,该类本身是一个抽象类,不能够实例化,但它是 Adapter 类的基类,开发者只需要从该类继承,然后重写里面的抽象方法即可创建自定义的 Adapter。创建自定义 Adapter 必须实现以下几个方法。

- getView():返回列表中每一项显示的视图,可以是一个结构复杂的布局对应的 View,也可以是简单的 TextView 或 ImageView。
- getCount():返回列表中项的个数,根据这个值循环填充列表。
- getItemId():返回指定项的 ID。

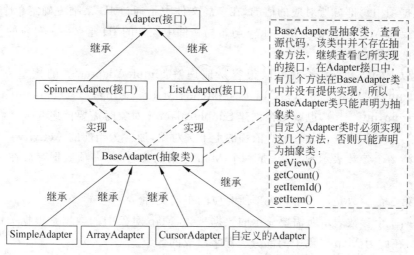

图 10-4　Adapter 层次结构图

- getItem()：返回指定项的对象。

其中，getView()方法和 getCount()方法最为关键，通过 getCount()方法即可知道列表中一共有多少项，然后通过 getCount()次循环调用 getView()方法获取每一项数据，并将每一项信息添加到 ListView 中形成包含数据的列表。

通过自定义 Adapter，可以使数据按我们想要的任何形式显示，还可以为项中的某一部分添加事件处理，缺点是代码比较多，需要重写多个方法。为此，Android 系统中提供了几个比较常用的 BaseAdapter 子类，例如 ArrayAdapter、SimpleAdapter、CursorAdapter 等，这些类各有自己的特点，适合一定的情况。例如，ArrayAdapter 适合于列表项中只包含文本的情况，SimpleAdapter 比较适合于列表项结构较为复杂的情况，CursorAdapter 比较适合于将数据库中查询的结果转换为列表的情况。

10.3.3　SimpleAdapter 介绍

SimpleAdapter 是一个简单且实用的 Adapter，它可以将静态的数据关联到 XML 布局文件中的某个 View 控件上。通常将静态的数据保存在 Map 对象的集合中，一个 Map 对象对应列表中的一项所包含的所有数据，通过 Map 对象中的关键字来区分一项中的每一类数据。例如，本案例中的一个景点包含 4 个部分数据，即景点图片、景点名称、景点简介、景点详情，因此 Map 对象中包含 4 个关键字，见代码的第 26～35 行，所有 Map 对象的集合即组成了整个数据源。

有了数据源以后，这些数据又应该如何显示呢？数据一般都要显示在相应的控件上，因此需要为列表中的单个项定义一个布局文件，布局文件可以非常简单，例如只包含一个控件，也可以非常复杂，由多种布局嵌套而成。在本案例中，每一项需要显示 3 个部分信息，布局中需要包含 3 个控件，并为每一个控件指定 ID 属性，代码见 item.xml 文件。

有了数据源和相应的控件之后，下面的关键在于如何将数据源中的数据项与布局文件中的控件进行关联，且不至于使数据显示混乱。在 Map 对象中是根据关键字 key 来唯一确定数据项中每部分所对应的值的,在布局文件中是通过 ID 来唯一确定单个控件的,因此只需要将 Map 对象的关键字 key 与布局文件中控件的 ID 建立一一对应的关系即可保证数据的一致。

有了以上的分析,读者就可以很容易地理解 SimpleAdapter 类的构造方法了,即 SimpleAdapter(Context context, List<?extends Map<String,?>>data,int resource, String[] from,int[] to),也就是在创建 SimpleAdapter 对象时需要传递 5 个参数。

(1) 第 1 个参数是 Adapter 所依赖的上下文对象,通常为当前的 Activity。

(2) 第 2 个参数表示数据源,是一个 Map 对象的集合,每一项数据存放在一个 Map 对象中。

(3) 第 3 个参数是每一个列表项所对应的布局文件。

(4) 第 4 个参数表示所需要显示的数据来源,Map 对象中的数据是根据关键字来唯一确定的,在此为 Map 对象中的部分关键字组成的数组。

(5) 第 5 个参数表示每部分数据如何显示,即指定显示的控件,而控件是根据 ID 来确定的,因此传递的是控件 ID 组成的数组。

注意:第 4 个参数与第 5 个参数之间存在一一对应的关系,根据第 4 个参数获取的数据将会在第 5 个参数指定的控件中显示,并且第 5 个参数中的元素必须在第 3 个参数指定的布局文件中。

10.3.4 ClipDrawable 介绍

ClipDrawable 表示从某个位图上截取的一个"图片片段",通过 ClipDrawable 实现图片徐徐展开效果的原理是不断地重复截取同一张图片,只是每次截取的片段的大小不同,最开始时截取的片段很小,然后不断增大,直到截取整个图片,给人的感觉就是徐徐展开。

ClipDrawable 对象既可以在代码中创建,也可以在 XML 文件中定义。在 XML 文件中定义 ClipDrawable 对象时,使用<clip.../>元素作为根元素,主要包含以下几个属性。

- android:drawable:指定需要截取的图片对象。
- android:clipOrientation:指定截取的方向,只有水平和垂直两个值。
- android:gravity:指定截取的对齐方式,例如从左到右、从右到左或者从中间向两边等。

直接在代码中创建 ClipDrawable 对象时,ClipDrawable(Drawable drawable, int gravity, int orientation)需要传递 3 个参数,第 1 个参数表示需要截取的图片对象;第 2 个参数为截取时的对齐方式是从左到右还是从右到左等;第 3 个参数为截取的方向是水平还是垂直。

在使用 ClipDrawable 对象截取图片时,是通过 setLevel(int level)方法来设置截取的区域大小,当 level 的值为 0 时,截取的区域为空;当 level 为 10 000 时,截取的就是整张图片了。本案例中是启动一个线程判断是否截取了整张图片,如果没有,则每隔 0.3s 发送一次请求,截取图片,每次截取时,使 level 的值递增 400。

10.4 知识扩展

10.4.1 raw 目录介绍

本案例中由于景点的介绍文字比较多,不宜放在代码中,而将每一个景点的介绍单独放在一个 TXT 文件中,然后在 res 目录下创建 raw 文件夹,再将这些文件放到 raw 文件夹下。

raw 文件夹是 Android 中存放原始资源文件的文件夹,该文件夹内的文件会原封不动地存储到设备上,不会被编译成二进制形式,可通过 R.raw.XXX 引用文件,使用 getResource().OpenRawResources(R.raw.XXX) 打开输入流,然后即可读取文件内容。在该文件夹下不能任意创建子文件夹。

在 Android 应用程序中,assets 文件夹以及 res 文件夹都是用于存放资源的,那么它们之间有什么区别呢?

- **assets**:用于存放需要打包到安装程序中的静态文件,存放在这里的资源都会原封不动地保存在安装包中,不会被编译成二进制形式。与 res 不同的是,assets 支持任意深度的子目录(即在该文件夹下可以任意创建子文件夹)。这些文件不会生成任何资源标记,必须使用以 /assets 开始(但不包含它)的相对路径名,需要使用 AssetManager 类访问,通过文件流的方式进行读取。
- **res**:用于存放应用程序的资源(如图标、界面布局等),会在 R.java 文件中生成标记,这里的资源会在打包时判断是否被使用,未使用的资源将不会打包到安装包中。该文件夹下包括一些固定的子文件夹,但不能任意创建子文件夹。

assets 文件夹、res 文件夹以及 raw 文件夹都可以存放资源,它们之间的区别见表 10-1。

表 10-1 assets、res、res/raw 文件夹对比

比 较 项	assets 文件夹	res 文件夹	res/raw 文件夹
是否在 R.java 中生成资源标记	否	是	是
是否能任意创建子文件夹	能	不能	不能
是否会被编译成二进制形式	不会	会	不会
是否完全打包到安装文件中	是	需判断	需判断
访问方式	AssetManager 类,通过文件流读取	R.XX.XXX 引用,通过 Resource 类的相应方法读取	R.raw.XXX 引用,通过 Resource 类的相应方法读取

10.4.2 Activity 概述

Activity 是 Android 的一种应用程序组件,该组件为用户提供了一个屏幕,用户在这个屏幕上进行操作即可完成一定的功能,例如打电话、拍照、发送邮件或查看地图等。

一个 Android 应用程序通常是由多个 Activity 组成的，但有且只有一个 Activity 被指定为主 Activity。当应用程序第一次启动时，系统会自动运行主 Activity。每个 Activity 都可以启动其他的 Activity 用于执行不同的功能。当一个新的 Activity 启动时，之前的那个 Activity 就会停止，但是系统会在堆栈中保存该 Activity。当用户使用完当前的 Activity 并按 Back 键时，该 Activity 将从堆栈中取出并销毁，然后之前的那个 Activity 将恢复并获取焦点。

如果想创建自己的 Activity，必须继承 Activity 基类或者是已存在的 Activity 子类，例如 ListActivity 等。在用户自己的 Activity 中可实现系统 Activity 类中所定义的一些回调方法，这些回调方法在用户的 Activity 状态发生变化时会由系统自动调用，其中最重要的回调方法就是 onCreate()。

当 Activity 被创建时，系统将会自动回调它的 onCreate() 方法，在该方法的实现中应该初始化一些关键的界面控件，最重要的是调用 Activity 的 setContentView() 方法来设置自己的 Activity 所对应的界面布局文件。为了管理应用程序界面中的各个控件，可调用 Activity 的 findViewById(int id) 方法来获取界面中的控件，然后即可修改该控件的属性和调用该控件的方法。

在定义好自己的 Activity 后，此时系统还不能访问该 Activity，如果想让系统访问，则必须在 AndroidManifest.xml 文件中进行注册、配置。在前面所写的程序中，我们也有自己的 Activity，但我们并没有对它进行配置，不是也可以访问吗？这是因为，在前面所有的程序中都只有一个 Activity，我们的开发工具在创建时自动地为它进行了配置，并把它作为主 Activity，默认配置如下：

```
1   <activity
2       android:name="iet.jxufe.cn.android.sceneryinfo.MainActivity"
3       android:label="@string/app_name"
4       android:theme="@android:style/Theme.Holo.Light.NoActionBar">
5       <intent-filter>
6           <action android:name="android.intent.action.MAIN"/>
7           <category android:name="android.intent.category.LAUNCHER"/>
8       </intent-filter>
9   </activity>
```

其中，<intent-filter>和</intent-filter>之间的内容表示该 Activity 是入口 Activity。

在配置自定义的 Activity 时，只需要在<application.../>标签内添加<activity.../>子标签，然后设置其相关属性即可，属性主要有以下几个。

- **name**：指定 Activity 实现类的类名，需要用完整的包名＋类名。
- **icon**：指定该 Activity 对应的图标，显示在 Activity 的标题栏上，一般不用设置。
- **label**：指定该 Activity 的文字标签，也显示在标题栏上。

此外，在配置 Activity 时通常还可以指定一个或多个<intent-filter.../>元素，该元素用于指定该 Activity 可响应的 Intent 的条件。

在上述配置中,只有 name 属性是必需的,其他属性或标签元素都是可选的。

Activity 在创建和配置完毕后,即可通过 **startActivity**(Intent intent)方法启动 Activity,该方法需要传入一个 Intent 类型的参数,该参数是对用户所需要启动的 Activity 的描述,既可以是一个确切的 Activity 类,也可以是所需要启动的 Activity 的一些特征,然后由系统查找符合该特征的 Activity,如果有多个 Activity 符合该要求,系统将会以下拉列表的形式列出所有的 Activity,然后由用户选择具体启动哪一个。

在本案例中通过 intent.setClass(MainActivity.this, SceneryShowActivity.class)明确指定需要启动的 Activity 类,同时可以向启动的 Activity 传递一些数据,数据存放在 Intent 对象中。该对象通过键值对的形式保存数据,在目标 Activity 中,可以通过相关的键获取相应的值。

10.5 思考与练习

(1) 本案例中图片展开的效果是水平的,从中间开始向两边慢慢展开,请尝试修改代码,使其从上到下慢慢展开。

(2) 为下拉列表自定义 Adapter,在写一个类继承自 BaseAdapter 时必须重写父类中的一些方法,以下(　　)方法不是必需的。

  A) getCount()　　　　　　　　　B) getView()
  C) getItem()　　　　　　　　　 D) getDropDownView()

(3) 在 Android 中包含了很多 Adapter 的相关类,下列选项中(　　)类不是从 BaseAdapter 继承而来的。

  A) ArrayAdapter　　　　　　　　B) SimpleAdapter
  C) CursorAdapter　　　　　　　 D) PagerAdapter

(4) 以下关于 SimpleAdapter 构造方法中参数的描述不正确的是(　　)。

  A) 第 1 个参数为 Context 上下文对象,通常只需要传入当前的 Activity 对象
  B) 第 2 个参数为列表的数据来源,既可以是一个数组,也可以是一个集合
  C) 第 3 个参数为列表中每一项的布局文件,该布局中可以包含多个控件
  D) 第 4 个参数与第 5 个参数之间存在一一对应的关系,根据第 4 个参数获取的
    数据将会在第 5 个参数所指定的控件中显示,并且第 5 个参数中的元素必须
    在第 3 个参数指定的布局文件中

(5) 在配置 Activity 时,下列选项中必不可少的是(　　)。

  A) android:name　　　　　　　　B) <action.../>
  C) <intent-filter.../>　　　　　　　D) <category.../>

第11章

财大新闻——ListView 延迟加载效果

11.1 案例概述

本案例主要介绍 ListView 控件的使用,用列表来显示财大新闻信息,与前面南昌景点信息的显示非常类似,每条新闻包含图片、标题、简单介绍 3 个部分。不同之处有两个:①所有的新闻信息保存在数据库中,而不是直接在代码中指定;②新闻列表中并不是立即显示所有的新闻数据,而是只显示部分新闻记录,当用户滑动到最后一条新闻时,显示正在加载信息,然后加载新的记录,并添加到列表中。本案例涉及复杂列表项的构建、数据库的操作、滚动事件处理等,程序运行效果如图 11-1 至图 11-3 所示。

图 11-1 程序运行效果图 1

图 11-2 程序运行效果图 2

图 11-3 程序运行效果图 3

11.2 关键代码

主界面布局文件:11\ListViewDelayLoad\res\layout\activity_main.xml

```
1    <LinearLayout xmlns:android="http://schemas.android.com/apk/res/android"
2        xmlns:tools="http://schemas.android.com/tools"
3        android:layout_width="match_parent"
```

第11章 财大新闻——ListView 延迟加载效果

```xml
 4        android:layout_height="match_parent"
 5        android:background="#ccbbaa"
 6        android:orientation="vertical" >
 7        <TextView
 8            android:layout_width="wrap_content"
 9            android:layout_height="wrap_content"
10            android:layout_gravity="center_horizontal"
11            android:padding="10dp"
12            android:text="@string/title"
13            android:textColor="#000000"
14            android:gravity="center_vertical"
15            android:drawableLeft="@drawable/logo"
16            android:textSize="24sp" />
17        <ListView
18            android:id="@+id/news"
19            android:layout_width="match_parent"
20            android:layout_height="wrap_content"
21            android:divider="#aaaaaa"
22            android:dividerHeight="2dp"
23            android:background="#aabbcc"
24            android:gravity="center" />
25    </LinearLayout>
```

新闻列表中每一项的布局文件：11\ListViewDelayLoad\res\layout\item.xml

```xml
 1    <LinearLayout xmlns:android="http://schemas.android.com/apk/res/android"
 2        xmlns:tools="http://schemas.android.com/tools"
 3        android:layout_width="match_parent"
 4        android:layout_height="wrap_content"
 5        android:layout_margin="10dp"
 6        android:orientation="horizontal" >
 7        <ImageView
 8            android:id="@+id/image"
 9            android:layout_width="75dp"
10            android:layout_height="60dp"
11            android:scaleType="fitXY"
12            android:layout_margin="10dp"
13            android:contentDescription="@string/imgInfo" />
14        <LinearLayout
15            android:layout_width="match_parent"
16            android:layout_height="wrap_content"
17            android:layout_marginTop="5dp"
18            android:paddingRight="5dp"
19            android:orientation="vertical" >
```

```
20        <TextView
21            android:id="@+id/name"
22            android:layout_width="match_parent"
23            android:layout_height="wrap_content"
24            android:textColor="#000000"
25            android:singleLine="true"
26            android:ellipsize="end"
27            android:textSize="18sp" />
28        <TextView
29            android:id="@+id/info"
30            android:layout_width="wrap_content"
31            android:layout_height="wrap_content"
32            android:gravity="left"
33            android:textColor="#0000ee"
34            android:textSize="14sp"
35            android:maxLines="2"
36            android:layout_marginTop="5dp"
37            android:ellipsize="end"/>
38    </LinearLayout>
39 </LinearLayout>
```

显示底部加载信息的布局文件：11\ListViewDelayLoad\res\layout\load.xml

```
1  <LinearLayout xmlns:android="http://schemas.android.com/apk/res/android"
2      xmlns:tools="http://schemas.android.com/tools"
3      android:layout_width="match_parent"
4      android:layout_height="match_parent"
5      android:gravity="center"
6      android:orientation="horizontal" >
7      <ProgressBar
8          android:layout_width="wrap_content"
9          android:layout_height="wrap_content" />
10     <TextView
11         android:text="@string/load"
12         android:textSize="20sp"
13         android:textColor="#0000ff"
14         android:gravity="center_vertical"
15         android:layout_width="wrap_content"
16         android:layout_height="wrap_content"/>
17 </LinearLayout>
```

程序代码：11\ListViewDelayLoad\src\iet\jxufe\cn\android\listviewdelayload\MainActivity.java

```
1  public class MainActivity extends Activity {
```

```
2       private ListView news;                    //显示新闻的列表
3       private SimpleAdapter simpleAdapter;//关联数据源的 Adapter
4       //保存已加载的新闻数据的集合
        private List<Map<String,Object>>newsList=new ArrayList<Map<String,
           Object>>();
5       private MyOpenHelper mHelper;             //数据库辅助类
6       private SQLiteDatabase mDB;               //数据库封装类
7       private int totalCount;                   //新闻的总条数,包括加载的和未加载的
8       private int loadedCount;                  //已加载的新闻条数
9       private int lastItem;                     //显示可见的最后一项的序号
10      private View footView;                    //显示底部加载信息的控件
11      private int loadItemNum=4;                //每次加载的项数
12      private Handler myHandler=new Handler();
13      protected void onCreate(Bundle savedInstanceState){
14          super.onCreate(savedInstanceState);
15          requestWindowFeature(Window.FEATURE_NO_TITLE); //不显示标题栏
16          setContentView(R.layout.activity_main);
17          mHelper=new MyOpenHelper(this,"news.db",null,1);
18          mDB=mHelper.getReadableDatabase();                //获取数据库
19          news=(ListView)findViewById(R.id.news);
20          //将布局文件转换成 View 控件,并添加到列表中
            footView=getLayoutInflater().inflate(R.layout.load,null);
21          news.addFooterView(footView);
22          newsList=getData("select * from news_tb order by _id limit 0,6",null);
23          simpleAdapter=new SimpleAdapter(this,newsList,R.layout.item,
               new String[] { "image","title","info" },new int[] {
               R.id.image,R.id.name,R.id.info });
24          news.setAdapter(simpleAdapter);
25          totalCount=getCount();
26          MyOnScrollListener myListener=new MyOnScrollListener();
            //创建滚动事件监听器
27          news.setOnScrollListener(myListener);   //为列表添加滚动事件监听器
28      }
29      private class MyOnScrollListener implements OnScrollListener {
30          public void onScroll(AbsListView view,int firstVisibleItem,
31              int visibleItemCount,int totalItemCount){
32              lastItem=firstVisibleItem+ visibleItemCount-1;
33              loadedCount=simpleAdapter.getCount();       //记录已加载的项
34              if(loadedCount>=totalCount){
                //如果所有项都加载完,不再显示加载信息
35                  news.removeFooterView(footView);
36              }
37          }
38          public void onScrollStateChanged(AbsListView view,int scrollState){
```

```
39                    //注意,footView也算一项,当显示footView时
                      //lastItem的值与loadedCount的值相等
40              if(lastItem==loadedCount
41                  && scrollState==OnScrollListener.SCROLL_STATE_IDLE){
42                  if(loadedCount<=totalCount){    //如果没有完全加载完,则记载新项
43                      myHandler.postDelayed(new Runnable(){  //延迟3s执行
44                          public void run(){
45                              if(loadedCount+ loadItemNum>totalCount){
46                                  loadMore(totalCount-loadedCount);
                                    //加载最后几项数据
47                              } else {
48                                  loadMore(loadItemNum);     //加载默认多个数据
49                              }
50                          }
51                      },3000);
52                  }
53              }
54          }
55      }
56      public void loadMore(int num){
57          List<Map<String,Object>>list=getData(
                "select * from news_tb order by _id limit ?,?",new String[]{
                    loadedCount+ "",num+ "" });
58          newsList.addAll(list);              //将新加载的新闻添加到新闻列表中
59          simpleAdapter.notifyDataSetChanged();       //更新列表显示
60      }
61      public List<Map<String,Object>>getData(String sql,String[] args){
        //根据查询语句获取新闻数据
62          List<Map<String,Object>>list=new ArrayList<Map<String,Object>>();
63          Cursor cursor=mDB.rawQuery(sql,args);       //查询符合条件的记录
64          while(cursor.moveToNext()){
            //循环遍历每条记录,获取相应数据,并将数据保存到集合中
65              Map<String,Object>item=new HashMap<String,Object>();
66              item.put("image",cursor.getString(cursor.getColumnIndex
                    ("image")));
67              item.put("title",cursor.getString(cursor.getColumnIndex
                    ("title")));
68              item.put("info",cursor.getString(cursor.getColumnIndex
                    ("info")));
69              list.add(item);
70          }
71          return list;
72      }
73      //查询数据库中一共有多少条记录
```

```
74      public int getCount(){
75          Cursor cursor=mDB.rawQuery("select count(*)from news_tb",null);
76          if(cursor.moveToNext()){
77              return cursor.getInt(0);
78          }
79          return 0;
80      }
81      protected void onDestroy(){
82          if(mDB!=null){
83              mDB.close();
84          }
85          super.onDestroy();
86      }
}
```

数据库辅助类：11\ListViewDelayLoad\src\iet\jxufe\cn\android\listviewdelayload\MyOpenHelper.java

```
1   public class MyOpenHelper extends SQLiteOpenHelper {
2       public String createTableSQL="create table if not exists news_tb" +
3           "(_id integer primary key autoincrement,image,title,info)";     //建表语句
4       public MyOpenHelper(Context context,String name,CursorFactory factory,
            int version){
5           super(context,name,factory,version);
6       }
7       //创建数据库后回调该方法,执行建表操作和插入初始化数据
        public void onCreate(SQLiteDatabase db){
8           db.execSQL(createTableSQL);
9           //执行初始化插入语句,在此省略
10      }
11      //数据库版本更新时回调该方法
        public void onUpgrade(SQLiteDatabase db,int oldVersion,int newVersion){
12          System.out.println("版本变化:"+oldVersion+"-------->"+newVersion);
13      }
14  }
```

11.3 代码分析

11.3.1 ListView 延迟加载原理

为了提高 ListView 的效率和应用程序的性能，对于列表项比较多的 ListView，Android 应用程序通常不会一次性加载 ListView 的所有项信息，而是采取分批加载策略，随着用户的滑动，动态地从后台加载所需的数据，并添加到 ListView 控件中，这样可以极大地改善应用程序的性能和用户体验。

具体操作：ListView 初始化时预加载 N 条记录,当用户滑动 ListView 控件到最后一条记录时显示加载提示信息,然后从后台加载 M 条数据,加载完毕后更新 ListVIew。在这个过程中需要定义以下几个变量。

- totalCount：数据库中包含的所有新闻的数目。
- loadedCount：列表中已经加载的新闻的数目。
- loadItemNum：每次加载时加载的新闻数目。
- lastItem：列表中可见的最后一项的序号。

本案例中显示加载提示信息的控件并不直接显示在屏幕的底部,而是随着 ListView 的滚动显示的,位于 ListView 的最后,属于 ListView 的一部分,实际上它也是作为 ListView 中的一项。只是它比较特殊,它的结构与其他普通项不同,不管何时添加到 ListView 中,它永远是 ListView 的最后一项,可通过 ListView 的 addFooterView() 方法添加到 ListView 中。

ListView 延迟加载的本质就是根据 ListView 滚动条的状态来决定是否额外加载数据,因此需要监听 ListView 的滚动状态,为 ListView 添加滚动事件监听器,实现该监听器需要重写其中的两个方法,即 onScroll() 方法和 onScrollStateChanged() 方法。

onScroll（AbsListView view，int firstVisibleItem，int visibleItemCount，int totalItemCount)方法用于监听滚动事件,该方法中包含 4 个参数,其中 view 表示滚动的控件,在此为 ListView；firstVisibleItem 表示当时该控件上用户可见的第一项的序号；visibleItemCount 表示该控件上用户可见项的数目；totalItemCount 表示该控件上当时一共有多少项,包括底部显示加载提示信息的项。lastItem 与 firstVisibleItem、VisibleItemCount 之间的数量关系是 lastItem＝firstVisibleItem＋visibleItemCount－1,减 1 主要是因为序号从 0 开始。当所有的记录都加载完毕时,就不需要显示底部的正在加载提示信息了,所以需要在 onScroll() 方法中判断 loadedCount 是否大于等于 totalCount 的值。代码见第 **34～36** 行。

onScrollStateChanged(AbsListView view，int scrollState)方法用于监听滚动条的状态发生变化事件,该方法包含两个参数,其中,view 表示滚动的控件,在此为 ListView；scrollState 表示滚动条的状态,在此主要是判断滚动条是否不能再滚动了,即滚动条的状态为 OnScrollListener.SCROLL_STATE_IDLE。滚动条不能再滚动有两种情况,一是向下滚动到最后一项；二是向上滚动到第一项。我们需要处理的是第一种,所以还需要判断 lastItem 是否等于 loadedCount,然后判断是否还有数据没有加载完,如果还有,即 loadedCount＜totalCount,则需要加载数据,那么具体加载多少项呢？默认情况下每次加载 4 项,如果最后不足 4 项,则加载所有剩余项。代码见第 **40～53** 行。

11.3.2 SQLite 数据库介绍

SQLite 数据库是 Android 中内嵌的轻量级关系型数据库,它不像 Oracle、MySQL、SQLServer 等大型数据库那样需要安装、启动服务进程。SQLite 数据库本质上只是一个文件,它支持绝大部分的 SQL 语法,非常适用于资源有限的设备上的适量数据的存取。它包含了操作本地数据的所有功能,简单易用、反应快。

SQLite 内部只支持 NULL、INTEGER、REAL、TEXT 和 BLOB 这 5 种数据类型，在 SQLite 中可以把各种类型的数据保存到任何字段中，而不用关心字段声明的数据类型是什么。例如，可以把字符串类型的值存入 INTEGER 类型字段中。因此在编写建表语句时，可以省略数据列后面的类型声明。例如，本案例中的建表语句为 create table if not exists news_tb (_id integer primary key autoincrement, image, title, info)。但有一种情况例外，定义为 **INTEGER PRIMARY KEY** 的字段只能存储 **64** 位整数，当向这种字段保存除整数以外的数据时，**SQLite 会产生错误**。

SQLite 数据库支持绝大部分 SQL 92 语法，也允许开发者使用 SQL 语句操作数据库中的数据。常见的 SQL 标准语句示例如下：

1. 查询

`select * from 表名 where 条件子句 group by 分组子句 having ... order by 排序子句`

例如：

`select * from person order by id desc`

查询 person 表中所有的记录，按 ID 号降序排列

`select name from person group by name having count(*)>1`

查询 person 表中 name 字段值出现超过 1 次的 name 字段的值

2. 分页

`select * from 表名 limit 跳过的记录数,显示的记录数`

例如：

`select * from person limit 3,5`

从 person 表中获取 5 条记录，跳过前面 3 条记录

3. 插入

`insert into 表名(字段列表) values(值列表)`

例如：

`insert into person(name, age) values('张三',26)`

向 person 表中插入一条记录，姓名为张三，年龄为 26 岁。

4. 更新

`update 表名 set 字段名=值 where 条件子句`

例如：

`update person set name='李四' where id=10`

将 ID 为 10 的记录的姓名改为李四。

5. 删除

`delete from 表名 where 条件子句`

例如：

`delete from person where id=10`

删除 person 表中 ID 为 10 的记录。

为了操作和管理 SQLite 数据库,在 Android 系统中提供了一些相关类,常用的有 SQLiteOpenHelper、SQLiteDatabase、Cursor,对于其他的用户可查看 Android 帮助文档中的 android.database.sqlite 包和 android.database 包。

SQLiteOpenHelper 是 Android 提供的管理数据库的工具类,主要用于数据库的创建、打开和版本更新等。该类是一个抽象类,在使用时需要创建 SQLiteOpenHelper 类的子类,并重写它的 onCreate()和 onUpgrade()方法(**这两个方法是抽象的,必须扩展**)。

在 SQLiteOpenHelper 类中包含的方法主要如下。

- SQLiteDatabase **getReadableDatabase**():以读写的方式打开数据库对应的 SQLite-Database 对象,该方法内部调用 getWritableDatabase()方法,返回的对象与 getWritableDatabase()方法返回的对象一致,当数据库磁盘空间满时,通过 getWritable-Database()方法打开数据库就会出错,这时 getReadableDatabase()方法会继续尝试以只读方式打开数据库。
- SQLiteDatabase **getWritableDatabase**():以写的方式打开数据库对应的 SQLite-Database 对象,一旦打开成功,将会缓存该数据库对象。
- abstract void **onCreate**(SQLiteDatabase db):当数据库第一次被创建时调用该方法,通常在该方法中执行初始化操作,如创建表结构、插入初始数据等。
- abstract void **onUpgrade**(SQLiteDatabase db, int oldVersion, int newVersion):当数据库版本发生变化时调用该方法。
- void **onOpen**(SQLiteDatabase db):当数据库打开时调用该方法。

当调用 SQLiteOpenHelper 的 getWritableDatabase()或者 getReadableDatabase()方法获取 SQLiteDatabase 实例的时候,系统根据数据库名来判断该数据库是否存在,如果数据库不存在,Android 系统会自动生成一个数据库,然后回调 onCreate()方法,在 onCreate()方法中执行建表语句及添加一些应用需要的初始化数据。如果数据库存在,系统再根据数据库的版本号来判断是否需要更新,如果版本与之前的不一致,则需要更新,自动调用 onUpgrade()方法,在该方法中根据业务需要执行数据表结构及数据更新。

SQLiteDatabase 是 Android 提供的 SQLite 数据库的封装类,该类封装了一些操作数据库的 API,通过该类可以完成数据的添加(Create)、查询(Retrieve)、更新(Update)和删除(Delete)操作,分别提供了 insert()、query()、update()、delete()方法。除此之外,在 SQLiteDatabase 中还提供了两个非常实用的方法,即 execSQL()和 rawQuery()方法。execSQL()方法可以执行 insert、delete、update 和 create table 之类有更改行为的 SQL 语句,而 rawQuery()方法用于执行 select 语句。方法声明如下。

- execSQL(String sql,Object[] bindArgs)：执行带占位符的 SQL 语句，如果 SQL 语句中没有占位符，则第二个参数可传 null。
- execSQL(String sql)：执行 SQL 语句。
- rawQuery(String sql,String[] selectionArgs)：执行带占位符的 SQL 查询。

Cursor 接口主要用于存放查询记录的接口，Cursor 是结果集游标，用于对结果集进行随机访问，如果读者熟悉 JDBC，可发现 Cursor 与 JDBC 中的 ResultSet 作用很相似，提供了以下方法来移动查询结果的记录指针。

- move(int offset)：将记录指针向上或向下移动指定的行数。若 offset 为正数，向下移动，为负数，向上移动。
- moveToNext()方法：可以将游标从当前记录移动到下一记录，如果已经移过了结果集的最后一条记录，返回结果为 false，否则为 true。
- moveToPrevious()方法：用于将游标从当前记录移动到上一记录，如果已经移过了结果集的第一条记录，返回值为 false，否则为 true。
- moveToFirst()方法：用于将游标移动到结果集的第一条记录，如果结果集为空，返回值为 false，否则为 true。
- moveToLast()方法：用于将游标移动到结果集的最后一条记录，如果结果集为空，返回值为 false，否则为 true。

使用 SQLiteDatabase 进行数据库操作的步骤如下：

（1）创建数据库的辅助类对象，在此为 MyOpenHelper 类，指定数据库的名称和版本号。

（2）调用辅助类的 getReadableDatabase()或者 getWritableDatabase()方法，获取 SQLiteDatabase 对象，该对象代表了与数据库的连接。

（3）调用 SQLiteDatabase 对象的相关方法来执行增、删、查、改操作。

（4）对数据库操作的结果进行处理，例如判断是否插入、删除或者更新成功，将查询结果记录转化成以列表显示等。

（5）关闭数据库连接，回收资源。

SQLite 数据库本质上是一个文件，当数据库创建成功后，会在手机的/data/data/应用程序所在包/databases/文件夹下生成相应的数据库文件，用户可通过 DDMS 视图查看并导出该数据库，在本案例中数据库路径为 data/data/iet.jxufe.cn.android.listviewdelayload/databases/news.db，如图 11-4 所示。

图 11-4　数据库存储路径

在导出数据库后，可通过 Android SDK 的 tool 目录下提供的 sqlite3.exe 打开数据库文件，类似于 MySQL 提供的命令行窗口。如果用户将 tool 目录添加到了环境变量，则只需要通过命令行进入到数据库文件所在的目录,然后输入命令"sqlite3 数据库名"即可打开数据库。如果用户没有将该目录添加到环境变量,则需要进入 Android SDK 安装目录下的 tool 目录,然后输入命令"sqlite3 数据库所在目录绝对路径/数据库名"即可打开数据库,打开数据库后可执行相应的 SQL 语句进行增、删、查、改操作。

注意:通过命令行查看数据库内容时,中文在命令行上会显示乱码。

11.4 知识扩展

在开发包含数据库操作的应用时,如果对数据库辅助类中的 onCreate()方法进行了更改,例如数据库的建表语句或者初始化值有变化,在测试时,一定要先把手机中的数据库删除。否则由于手机上已经存在该数据库,系统不会重复调用 onCreate()方法,从而也就达不到更改目的。

在进行数据库操作时,通常需要根据传递的参数拼接 SQL 语句,例如有一个方法,用于根据 ID 从数据库中查询记录,在调用该方法时需要传递一个 ID,然后方法内部拼接 SQL 语句,执行查询操作。当参数比较多时,拼接 SQL 语句非常不方便,并且容易出错,也不安全。比较好的方法就是将需要动态变化的部分通过占位符来表示,在 SQL 语句中占位符通过?号来表示,然后对这些?号逐一进行赋值。SQLiteDatabase 中的 exeSQL()方法以及 rawQuery()方法都提供了执行带占位符的 SQL 语句的操作。在本案例中插入初始化值时就使用了占位符。例如"db.execSQL("insert into news_tb(image,title,info) values(?,?,?)",new String[]{R.drawable.news1+"","各地校友会和校友献礼母校 90 华诞","谁言寸草心,报得三春晖。近期,各地校友会和校友怀着对母校的感恩之情,纷纷捐赠纪念品,献礼母校 90 华诞..."});"。

11.5 思考与练习

(1) 数据库的创建过程是怎样的?如果在定义辅助类时在其 onCreate()方法中执行了建表语句,当还未创建数据库时直接查找记录会不会出错?

(2) 数据库的扩展名有要求吗?

(3) SQLite 允许把各种类型的数据保存到任何类型字段中,开发者可以不用关心声明该字段所使用的数据类型,这种说法对吗?

(4) 在以下数据类型中,不是 SQLite 内部支持的类型是(   )。

  A) BLOB   B) INTEGER

  C) VARCHAR   D) REAL

(5) 下列关于 SQLiteOpenHelper 的描述不正确的是(   )。

  A) SQLiteOpenHelper 是 Android 中提供的管理数据库的工具类,主要用于数据库的创建、打开、版本更新等,它是一个抽象类

  B) 继承 SQLiteOpenHelper 的类,必须重写它的 onCreate()方法

  C) 继承 SQLiteOpenHelper 的类,必须重写它的 onUpgrade()方法

D) 继承 SQLiteOpenHelper 的类，可以提供构造方法，也可以不提供构造方法

（6）SQLiteOpenHelper 是 Android 中提供的管理数据库的工具类，用于管理数据库的创建、版本更新、打开等，它是一个抽象类。如果创建一个该类的子类，在以下方法中，不是必须要包含在新创建的类中的方法是(　　)。

  A) 构造方法   B) onCreate()
  C) onUpgrade()   D) getReadableDatabase()

第12章

财大新闻——ListView 下拉刷新效果

12.1 案例概述

本案例主要介绍列表的下拉刷新效果,列表的内容与上一章一致。下拉刷新经常用于一些实时数据更新应用,如新闻客户端。当用户看新闻时,可能会有新的内容发布,而用户往往不希望实时收到新闻提示,而是希望自己在查看新闻时主动刷新页面。本案例就是通过用户从顶部往下拉列表来刷新新闻,将最新的新闻显示在最上方,程序运行效果如图 12-1 至图 12-5 所示。

图 12-1　程序运行效果图 1

图 12-2　程序运行效果图 2

图 12-3　程序运行效果图 3

图 12-4　程序运行效果图 4

图 12-5　程序运行效果图 5

12.2 关键代码

主界面布局文件：12\DropDownRefresh\res\layout\activity_main.xml

```
1  <LinearLayout xmlns:android="http://schemas.android.com/apk/res/android"
2      xmlns:tools="http://schemas.android.com/tools"
3      android:layout_width="match_parent"
4      android:layout_height="match_parent"
5      android:background="#ccbbaa"
6      android:orientation="vertical" >
7      <TextView
8          android:layout_width="wrap_content"
9          android:layout_height="wrap_content"
10         android:layout_gravity="center_horizontal"
11         android:drawableLeft="@drawable/logo"
12         android:gravity="center_vertical"
13         android:padding="10dp"
14         android:text="@string/title"
15         android:textColor="#000000"
16         android:textSize="24sp" />
17     <iet.jxufe.cn.android.dropdownrefresh.RefreshListView
18         android:id="@+id/news"
19         android:layout_width="match_parent"
20         android:layout_height="wrap_content"
21         android:background="#aabbcc"
22         android:divider="#aaaaaa"
23         android:dividerHeight="2dp"/>
24 </LinearLayout>
```

新闻列表中每一项的布局文件：12\DropDownRefresh\res\layout\item.xml

```
1  <LinearLayout xmlns:android="http://schemas.android.com/apk/res/android"
2      xmlns:tools="http://schemas.android.com/tools"
3      android:layout_width="match_parent"
4      android:layout_height="wrap_content"
5      android:layout_margin="10dp"
6      android:orientation="horizontal" >
7      <ImageView
8          android:id="@+id/image"
9          android:layout_width="75dp"
10         android:layout_height="60dp"
11         android:scaleType="fitXY"
12         android:layout_margin="10dp"
```

```
13          android:contentDescription="@string/imgInfo" />
14      <LinearLayout
15          android:layout_width="match_parent"
16          android:layout_height="wrap_content"
17          android:layout_marginTop="5dp"
18          android:paddingRight="5dp"
19          android:orientation="vertical" >
20          <TextView
21              android:id="@+id/name"
22              android:layout_width="match_parent"
23              android:layout_height="wrap_content"
24              android:textColor="#000000"
25              android:singleLine="true"
26              android:ellipsize="end"
27              android:textSize="18sp" />
28          <TextView
29              android:id="@+id/info"
30              android:layout_width="wrap_content"
31              android:layout_height="wrap_content"
32              android:gravity="left"
33              android:textColor="#0000ee"
34              android:textSize="14sp"
35              android:maxLines="2"
36              android:layout_marginTop="5dp"
37              android:ellipsize="end"/>
38      </LinearLayout>
39  </LinearLayout>
```

显示顶部刷新信息的布局文件：12\DropDownRefresh\res\layout\refresh_header.xml

```
1   <?xml version="1.0" encoding="utf-8"?>
2   <LinearLayout xmlns:android="http://schemas.android.com/apk/res/android"
3       android:layout_width="match_parent"
4       android:layout_height="wrap_content"
5       android:gravity="center"
6       android:background="#aaaabb"
7       android:paddingBottom="5dp"
8       android:orientation="horizontal" >
9       <!--显示加载的进度条,默认是不可见的 -->
10      <ProgressBar
11          android:id="@+id/progressbar"
12          android:layout_width="wrap_content"
13          android:layout_height="wrap_content"
14          android:layout_gravity="center"
```

```
15          style="?android:attr/progressBarStyle"/>
16      <!--图片信息在下拉时显示向下的箭头,在释放时显示向上的箭头 -->
17      <ImageView
18          android:id="@+id/imageInfo"
19          android:layout_width="24dp"
20          android:layout_height="60dp"
21          android:layout_marginRight="5dp"
22          android:contentDescription="@string/imgInfo"/>
23      <!--头部提示信息 -->
24      <TextView
25          android:id="@+id/headerInfo"
26          android:layout_width="wrap_content"
27          android:layout_height="wrap_content"
28          android:gravity="center"
29          android:textColor="#0000ff"
30          android:textSize="14sp"
31          android:paddingTop="10dp"
32          android:paddingBottom="10dp"/>
33  </LinearLayout>
```

程序代码:12\DropDownRefresh\src\iet\jxufe\cn\android\dropdownrefresh\MainActivity.java

```
1  public class MainActivity extends Activity {
2      protected void onCreate(Bundle savedInstanceState){
3          super.onCreate(savedInstanceState);
4          requestWindowFeature(Window.FEATURE_NO_TITLE);   //不显示标题栏
5          setContentView(R.layout.activity_main);
6      }
7  }
```

数据库辅助类:12\DropDownRefresh\src\iet\jxufe\cn\android\dropdownrefresh\MyOpenHelper.java

```
1  public class MyOpenHelper extends SQLiteOpenHelper {
2      public String createTableSQL="create table if not exists news_tb" +
3          "(_id integer primary key autoincrement,image,title,info)"; //建表语句
4      public MyOpenHelper(Context context,String name,CursorFactory factory,
5              int version){
6          super(context,name,factory,version);
7      }
8      //数据库创建后回调该方法,执行建表操作和插入初始化数据的操作
       public void onCreate(SQLiteDatabase db){
9          db.execSQL(createTableSQL);
10         //执行初始化插入语句,在此省略
11     }
```

```
12      //数据库版本更新时回调该方法
        public void onUpgrade(SQLiteDatabase db,int oldVersion,int newVersion){
13          System.out.println("版本变化:"+oldVersion+"-------->"+newVersion);
14      }
15  }
```

自定义列表类:12\DropDownRefresh\src\iet\jxufe\cn\android\dropdownrefresh\MyOpenHelper.java

```
1   public class RefreshListView extends ListView {
2       private SimpleAdapter simpleAdapter;          //关联数据源的Adapter
3       private LinkedList<Map<String,Object>> newsList;  //保存已加载的新
                                                            闻数
                                                            //据的集合
4       private LinearLayout mHeaderView=null;        //顶部的布局
5       private TextView headerInfo;                  //顶部显示文本信息的控件
6       private ImageView imageInfo;                  //顶部显示图片信息的控件
7       private ProgressBar mProgressBar;             //顶部进度条
8       private int mHeaderHeight;                    //下拉列表头部View的高度
9       //private boolean hasNew;                     //是否包含新记录的标志
10      private int mCurrentScrollState;              //当前的滚动条状态
11      private MyOpenHelper mHelper;                 //数据库辅助类
12      private SQLiteDatabase mDB;                   //数据库封装类
13      //下拉刷新过程中自定义的几种状态
14      private final static int NONE_PULL_REFRESH=0;    //正常状态
15      private final static int ENTER_PULL_REFRESH=1;   //进入下拉刷新状态
16      private final static int OVER_PULL_REFRESH=2;    //进入松手刷新状态
17      private final static int EXIT_PULL_REFRESH=3;    //松手,反弹后加载状态
18      private int mPullRefreshState=0;              //记录刷新状态,默认为正常状态
19      private float mDownY;                         //按下时的坐标
20      private float mMoveY;                         //移动后的坐标
21      private SimpleDateFormat mSimpleDateFormat=new SimpleDateFormat(
            "yyyy-MM-dd hh:mm:ss");                   //时间的显示格式
22      //Handler处理消息时所包含的一些消息标志
23      private final static int REFRESH_BACKING=0;   //反弹中
24      private final static int REFRESH_BACKED=1;    //反弹结束后需要刷新
25      private final static int REFRESH_RETURN=2;    //反弹结束后不需要刷新
26      private final static int REFRESH_DONE=3;      //数据加载结束
27      public RefreshListView(Context context){      //构造方法,不包含属性
28          this(context,null);
29      }
30      public RefreshListView(Context context,AttributeSet attrs){
                                                        //构造方法,包含属性
31          super(context,attrs);
32          init(context);                            //进行初始化操作
```

第12章　财大新闻——ListView下拉刷新效果

```
33          setOnScrollListener(new MyScrollListener());    //添加滚动事件监听器
34          setSelection(1);                    //设置选中项,默认为0,头部View也算一项
35      }
36      public void init(final Context context){
37          mHelper=new MyOpenHelper(context,"news.db",null,1); //创建数据库辅助类
38          mDB=mHelper.getReadableDatabase();              //获取数据库
39          LayoutInflater layoutInflater=(LayoutInflater)context
                  .getSystemService(Context.LAYOUT_INFLATER_SERVICE);
              //获取布局填充器,用于将布局文件转换成View对象
40          mHeaderView=(LinearLayout)layoutInflater.inflate(
                  R.layout.refresh_header,null);      //将布局文件转换成View对象
41          addHeaderView(mHeaderView);             //将头部View添加到列表中
42          newsList=getData("select * from news_tb order by _id desc
                  limit 0,6",null);              //获取初始信息
43          simpleAdapter=new SimpleAdapter(context,newsList,R.layout.item,
                  new String[] {"image","title","info"},new int[] {
                      R.id.image,R.id.name,R.id.info});
44          this.setAdapter(simpleAdapter);
45          //根据ID找到布局中相应的控件
            headerInfo=(TextView)mHeaderView.findViewById(R.id.headerInfo);
46          imageInfo=(ImageView)mHeaderView.findViewById(R.id.imageInfo);
47          mProgressBar=(ProgressBar)mHeaderView.findViewById
                (R.id.progressbar);
48          reset();                            //初始状态
49          measureView(mHeaderView);               //指定头部控件的大小
50          mHeaderHeight=mHeaderView.getMeasuredHeight();  //获取头部View的高度
51      }
52      private void measureView(View child){
53          ViewGroup.LayoutParams p=child.getLayoutParams(); //获取控件的布局参数
54          if(p==null){
55              p=new ViewGroup.LayoutParams(ViewGroup.LayoutParams.
                  MATCH_PARENT,ViewGroup.LayoutParams.WRAP_CONTENT);
                  //默认宽度填充父容器,高度为内容包裹
56          }
57          int childWidthSpec=ViewGroup.getChildMeasureSpec(0,0,p.width);
            //获取宽度要求
58          int lpHeight=p.height;
59          int childHeightSpec;
60          if(lpHeight>0){
61              childHeightSpec=MeasureSpec.makeMeasureSpec(lpHeight,
                      MeasureSpec.EXACTLY);
                  //子控件的高度由父容器指定,是一个精确值,而不管子控件内容的多少
62          } else {
63              childHeightSpec=MeasureSpec.makeMeasureSpec(0,
```

```
                        MeasureSpec.UNSPECIFIED);
                        //子控件的高度不受限制,可以是任意值
64              }
65              child.measure(childWidthSpec,childHeightSpec); //子控件的宽度和高度
66          }
67          //根据查询语句获取新闻数据
            public LinkedList<Map<String,Object>>getData(String sql,String[] args){
68              LinkedList<Map<String,Object>>list=new LinkedList<Map<String,
                    Object>>();
69              Cursor cursor=mDB.rawQuery(sql,args);       //查询符合条件的记录
70              while(cursor.moveToNext()){
                //循环遍历每条记录,获取相应数据,并将数据保存到集合中
71                  Map<String,Object>item=new HashMap<String,Object>();
72                  item.put("image",cursor.getString(cursor.getColumnIndex
                        ("image")));
73                  item.put("title",cursor.getString(cursor.getColumnIndex
                        ("title")));
74                  item.put("info",cursor.getString(cursor.getColumnIndex
                        ("info")));
75                  list.add(item);
76              }
77              return list;
78          }
79          public class MyScrollListener implements OnScrollListener {
            //滚动事件监听器
80              public void onScrollStateChanged(AbsListView view,int scrollState){
81                  mCurrentScrollState=scrollState;        //记录当前的滚动条状态
82              }
83              public void onScroll(AbsListView view,int firstVisibleItem,
                    int visibleItemCount,int totalItemCount){   //滚动时调用
84                  if(mCurrentScrollState==SCROLL_STATE_TOUCH_SCROLL
                        && firstVisibleItem==0 &&mHeaderView.getBottom()>=0 &&
                        mHeaderView.getBottom()<mHeaderHeight)){
85                      //手指在滑动滚动条、头部View还没有完全显示出来时,进入且仅进入
                        //下拉刷新状态
86                      if(mPullRefreshState==NONE_PULL_REFRESH){
87                          mPullRefreshState=ENTER_PULL_REFRESH;
88                      }
89                  } else if(mCurrentScrollState==SCROLL_STATE_TOUCH_SCROLL
                        && firstVisibleItem==0
                        &&(mHeaderView.getBottom()>=mHeaderHeight)){
                    //滚动达到或超过头部View的高度时,进入松手刷新状态
90                      if(mPullRefreshState==ENTER_PULL_REFRESH
                            ||mPullRefreshState==NONE_PULL_REFRESH){
```

```
91              mPullRefreshState=OVER_PULL_REFRESH;
92              //下面是进入松手刷新状态需要做的一个显示改变
93              mDownY=mMoveY;                    //记录进入松手状态时的坐标
94              headerInfo.setText("松手即可刷新\n上次刷新时间:"
                    +mSimpleDateFormat.format(new Date()));
95              imageInfo.setImageResource(R.drawable.up_arrow);
96          }
97       } else if(mCurrentScrollState==SCROLL_STATE_TOUCH_SCROLL
              && firstVisibleItem !=0){
98          //当可见项不是第一项时,即还没有拖动到顶部
99          if(mPullRefreshState==ENTER_PULL_REFRESH){
100             mPullRefreshState=NONE_PULL_REFRESH;
101         }
102      } else if(mCurrentScrollState==SCROLL_STATE_FLING
103            && firstVisibleItem==0){
104         //飞滑状态,不能显示出 Header,也不能影响正常的飞滑
105         //只在正常情况下才恢复位置
106         if(mPullRefreshState==NONE_PULL_REFRESH){
107             setSelection(1);
108         }
109      }
110    }
111  }
112  public boolean onTouchEvent(MotionEvent ev){
113      switch(ev.getAction()){
114      case MotionEvent.ACTION_DOWN:              //记下按下位置,改变
115          mDownY=ev.getY();
116          break;
117      case MotionEvent.ACTION_MOVE:              //移动时手指的位置
118          mMoveY=ev.getY();
119          if(mPullRefreshState==OVER_PULL_REFRESH){
120             //注意下面的 mDownY 在 onScroll 的第二个 else 中被改变了
121             mHeaderView.setPadding(0,(int)((mMoveY-mDownY)/2),0,
                    mHeaderView.getPaddingBottom());  //上边距为拖动距离的1/2
122         }
123         break;
124      case MotionEvent.ACTION_UP:        //松手时将会回弹,并且隐藏头部 View
125         if(mPullRefreshState==OVER_PULL_REFRESH
126            || mPullRefreshState==ENTER_PULL_REFRESH){
127             final Timer timer=new Timer();
128             timer.schedule(new TimerTask(){
129                 public void run(){
130                     while(mHeaderView.getPaddingTop()>1){
131                         Message msg=mHandler.obtainMessage();
```

```
132                        msg.what=REFRESH_BACKING;        //发送反弹的消息
133                        mHandler.sendMessage(msg);
134                    }
135                    Message message=mHandler.obtainMessage();
136                    if(mPullRefreshState==OVER_PULL_REFRESH){
                            //如果原来是松开
137                        message.what=REFRESH_BACKED;
                            //反弹完成,加载数据
138                    } else {                            //如果原来只是下拉
139                        message.what=REFRESH_RETURN;
                            //反弹完成,不需要加载数据
140                    }
141                    mHandler.sendMessage(message);
142                    timer.cancel();
143                }
144            },0,100);
145        };
146        break;
147    }
148    return super.onTouchEvent(ev);
149 }
150 private Handler mHandler=new Handler(){
    //创建 Handler 对象,发送、接收和处理消息
151    public void handleMessage(Message msg){
152        switch(msg.what){
153        case REFRESH_BACKING:          //处理反弹,每次让上边距为前一次的 0.75
154            mHeaderView.setPadding(0,(int)(mHeaderView.getPaddingTop() *
                    0.75f),0,mHeaderView.getPaddingBottom());
155            break;
156        case REFRESH_BACKED:            //加载数据
157            imageInfo.setVisibility(View.GONE);
158            mProgressBar.setVisibility(View.VISIBLE);
159            headerInfo.setText("正在刷新...\n 上次刷新时间:"
160                    +mSimpleDateFormat.format(new Date()));
161            mHandler.postDelayed(new Runnable(){
162                public void run(){
163                    refreshing();        //执行刷新操作
164                }
165            },3000);
166            break;
167        case REFRESH_RETURN:            //反弹结束,不需要加载数据,恢复原状
168        case REFRESH_DONE:              //刷新结束后,恢复原始默认状态
169            reset();
170            break;
```

```
171            default:
172                break;
173         }
174      }
175   };
176   public void reset(){                       //恢复原来的状态
177      headerInfo.setText("下拉可以刷新\n上次刷新时间:"
178            +mSimpleDateFormat.format(new Date()));
179      mProgressBar.setVisibility(View.GONE);
180      imageInfo.setVisibility(View.VISIBLE);
181      imageInfo.setImageResource(R.drawable.down_arrow);
182      mHeaderView.setPadding(0,0,0,mHeaderView.getPaddingBottom());
183      mPullRefreshState=NONE_PULL_REFRESH;
184      setSelection(1);
185   }
186   public void refreshing(){
187      LinkedList<Map<String,Object>>list=getData(
188            "select * from news_tb order by _id desc limit ?,?",
189            new String[]{simpleAdapter.getCount()+"",2+""});
190      if(list !=null && list.size()!=0){
191         for(int j=0; j<list.size(); j++){
192            newsList.addFirst(list.get(j));
193         }
194         simpleAdapter.notifyDataSetChanged();    //更新列表显示
195         mPullRefreshState=EXIT_PULL_REFRESH;    //改变状态,发送消息
196         Message message=mHandler.obtainMessage();
197         message.what=REFRESH_DONE;
198         mHandler.sendMessage(message);
199      }
200   }
201 }
```

12.3 代码分析

ListView下拉刷新常用于一些数据实时更新的应用,每次刷新时都会发送请求到服务器,加载自上次刷新到现在这段时间内所生成的记录,如新浪微博客户端、网易新闻客户端等。下拉刷新的原理是在ListView的顶部添加一项,该项用于显示用户下拉操作的信息。默认情况下,该项不显示,即列表从第二项开始显示。当用户向下拉动滚动条时会根据第一项是否完全显示来提示用户操作,提示信息包括下拉可以刷新、松手即可刷新、正在刷新等。在此过程中涉及下面4种状态。

- NONE_PULL_REFRESH:正常状态,在该状态下看不见第一项,默认为该状态。
- ENTER_PULL_REFRESH:进入刷新状态,在该状态下,第一项会提示用户下

拉可以刷新信息,同时显示向下的箭头,该状态持续的时间为用户滑动滚动条,使第一项刚刚可以看见底部一直到第一项可以完全显示。

- OVER_PULL_REFRESH:进入松手刷新状态,在该状态下,第一项会提示用户松手即可刷新信息,同时显示向上的箭头,该状态持续的时间为第一项完全显示,用户继续向下滑动滚动条直到用户松手。
- EXIT_PULL_REFRESH:松手反弹状态,在该状态下,第一项提示系统正在刷新信息,同时显示刷新的进度条,用户松手时即进入该状态。

以上 4 种状态是根据用户对 ListView 列表的滚动状态的变化切换的,因此,需要为 ListView 添加滚动事件监听器,同时需要记录用户按下滚动条的坐标以及监听用户松开事件,所以还需要为 ListView 添加触摸事件。滚动事件监听器和触摸事件监听器代码见第 79~149 行。下拉刷新的主要流程包括以下几个内容。

(1) 下拉:从无到部分显示 ListView 的顶部 HeaderView,提示用户下拉可以刷新,此时刷新状态由 NONE_PULL_REFRESH 转为 ENTER_PULL_REFRESH。

(2) 继续下拉:从部分显示 HeaderView 到完全显示 HeaderView,达到刷新的最低高度要求,此时提示用户松手即可刷新,但效果仍允许用户继续下拉,主要是通过不断增大 HeaderView 的上边距,刷新状态由 ENTER_PULL_REFRESH 转为 OVER_PULL_REFRESH。

(3) 用户松手:可能用户下拉的远远不止 HeaderView 控件的高度,所以应先反弹回仅显示头部提示控件,主要是通过不断减小 HeaderView 的上边距达到反弹效果,然后提示用户正在刷新,刷新状态由 OVER_PULL_REFRESH 转为 EXIT_PULL_REFRESH。

(4) 刷新完成:刷新完成后,设置第二项为默认可见第一项,从而隐藏 HeaderView 控件,刷新状态由 EXIT_PULL_REFRESH 转为 NONE_PULL_REFRESH。

下拉刷新的主要流程如图 12-6 所示。

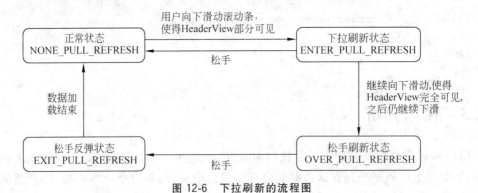

图 12-6 下拉刷新的流程图

12.4 知识扩展

在上一章延迟加载中,新闻数据保存在 ArrayList 集合中,延迟加载的数据添加到列表的最后,而本案例中新获取的数据需要添加到列表的开始,通过 ArrayList 不好实现,本案例中使用的是 LinkedList。ArrayList 与 LinkedList 都是 List 接口的实现类。ArrayList 底层是基于动态数组的数据结构,而 LinkedList 底层是基于链表的数据结构。它们在性能上各有优缺点,用户可根据具体情况选择使用。二者之间的关系如下:

(1) 对于 ArrayList 和 LinkedList 而言,在列表末尾增加一个元素所需要的开销都是固定的。对于 ArrayList 而言,主要是在内部数组中增加一项,指向所添加的元素,偶尔可能会导致对数组重新进行分配;而对于 LinkedList 而言,这个开销是统一的,分配一个内部 Entry 对象。

(2) 在 ArrayList 的中间插入或删除一个元素意味着这个列表中剩余的元素都会被移动;而在 LinkedList 的中间插入或删除一个元素的开销是固定的。

(3) LinkedList 不支持高效的随机元素访问。

(4) ArrayList 的空间浪费主要体现在列表的结尾预留一定的容量空间,而 LinkedList 的空间浪费体现在它的每一个元素都需要消耗相当的空间。

一般来说,如果操作是在列表数据的后面添加新项,而不是在前面或中间,并且需要随机地访问其中的元素时,使用 ArrayList 会比较方便;如果操作是在列表数据的前面或中间添加或删除数据,并且按照顺序访问其中的元素时,使用 LinkedList 比较方便。

在 LinkedList 类中提供了 addFirst()、addLast()方法,可以很方便地在列表的开始或结尾添加新的元素,非常符合本案例的需要,所以在本案例中使用 LinkedList。

12.5　思考与练习

请简单描述下拉刷新的过程,以及几种状态之间的转换关系。

学院介绍——选项卡切换效果

13.1 案例概述

本案例主要实现学院信息展示功能，通过不同的选项卡从多个方面介绍学院概况，并且在每个页面中都能够非常方便地切换到其他页面。本案例综合使用 TabHost 与 Fragment，整个应用中只包含一个主 Activity，在切换页面时动态地改变页面信息。TabHost 比较常见于具有导航功能的应用。在本案例中主要通过学院简介、现任领导、院属部门 3 个方面来介绍学院信息，程序运行效果如图 13-1 至图 13-3 所示。

图 13-1 学院简介页面

图 13-2 现任领导介绍页面

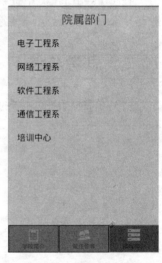
图 13-3 院属部门页面

13.2 关键代码

主界面布局文件：13\CollegeInfo\res\layout\activity_main.xml

```
1   <!--选项卡控件 -->
2   <TabHost xmlns:android="http://schemas.android.com/apk/res/android"
3       android:id="@+id/mTabHost"
4       android:layout_width="match_parent"
5       android:layout_height="match_parent"
```

```
 6        android:background="#aabbcc" >
 7        <!--整体放在垂直线性布局中,包括两部分选项卡和具体的页面显示 -->
 8        <LinearLayout
 9            android:layout_width="match_parent"
10            android:layout_height="match_parent"
11            android:orientation="vertical" >
12            <!--显示单个页面信息 -->
13            <FrameLayout
14                android:id="@android:id/tabcontent"
15                android:layout_width="match_parent"
16                android:layout_height="0dp"
17                android:layout_weight="1" >
18                <FrameLayout
19                    android:id="@+id/realcontent"
20                    android:layout_width="match_parent"
21                    android:layout_height="match_parent" />
22            </FrameLayout>
23            <!--底部显示所有的选项 -->
24            <TabWidget
25                android:id="@android:id/tabs"
26                android:layout_width="match_parent"
27                android:layout_height="wrap_content"
28                android:background="#66666666" >
29            </TabWidget>
30        </LinearLayout>
31  </TabHost>
```

每个选项卡包含两部分信息,即图标和标题,它们的布局信息如下。

单个选项布局文件:13\CollegeInfo\res\layout\tab.xml

```
 1  <LinearLayout xmlns:android="http://schemas.android.com/apk/res/android"
 2      xmlns:tools="http://schemas.android.com/tools"
 3      android:layout_width="match_parent"
 4      android:layout_height="match_parent"
 5      android:gravity="center_horizontal"
 6      android:background="@drawable/bg"
 7      android:padding="5dp"
 8      android:orientation="vertical">
 9      <ImageView
10          android:id="@+id/icon"
11          android:layout_width="30dp"
12          android:layout_height="30dp"
13          android:contentDescription="@string/imgInfo" />
14      <TextView
```

```
15          android:id="@+id/title"
16          android:textColor="#0000ff"
17          android:textSize="12sp"
18          android:layout_width="wrap_content"
19          android:layout_height="wrap_content" />
20  </LinearLayout>
```

选项卡的背景图片：13\CollegeInfo\res\drawable-hdpi\bg.xml

```
1  <?xml version="1.0" encoding="utf-8"?>
2  <selector xmlns:android="http://schemas.android.com/apk/res/android">
3      <!--在选中或单击状态下是一种图片,在其他状态下是另外一种图片-->
4      <item android:state_selected="true" android:drawable=
           "@drawable/bg_choosed"/>
5      <item android:state_pressed="true" android:drawable=
           "@drawable/bg_choosed"/>
6      <item android:drawable="@drawable/bg_unchoosed"/>
7  </selector>
```

自定义的两种背景图片对应的代码如下。

选中时的图片：13\CollegeInfo\res\drawable-hdpi\bg_choosed.xml

```
1  <?xml version="1.0" encoding="utf-8"?>
2  <shape xmlns:android="http://schemas.android.com/apk/res/android" >
3      <solid android:color="#554455"/>
4      <stroke android:width="1dp"
5          android:color="#aa0000"/>
6  </shape>
```

未选中时的图片：13\CollegeInfo\res\drawable-hdpi\ bg_unchoosed.xml

```
1  <?xml version="1.0" encoding="utf-8"?>
2  <shape xmlns:android="http://schemas.android.com/apk/res/android" >
3      <solid android:color="#666666"/>
4      <stroke android:width="1dp"
5          android:color="#000055"/>
6  </shape>
```

学院介绍页面对应的布局文件以及主程序代码如下。

学院介绍页面布局文件：13\CollegeInfo\res\layout\college.xml

```
1  <LinearLayout xmlns:android="http://schemas.android.com/apk/res/android"
2      android:layout_width="match_parent"
3      android:layout_height="match_parent"
```

```xml
4      android:background="#aabbcc"
5      android:orientation="vertical">
6      <TextView
7          android:layout_width="match_parent"
8          android:layout_height="wrap_content"
9          android:gravity="center"
10         android:textSize="24sp"
11         android:background="#77ccbbaa"
12         android:padding="10dp"
13         android:text="@string/collegeTitle"/>
14     <ScrollView
15         android:layout_width="match_parent"
16         android:layout_height="wrap_content" >
17         <TextView
18             android:id="@+id/infoView"
19             android:textSize="16sp"
20             android:textColor="#004400"
21             android:layout_width="match_parent"
22             android:layout_height="wrap_content"
23             android:padding="5dp"/>
24     </ScrollView>
25 </LinearLayout>
```

页面主程序：13\CollegeInfo\src\iet\jxufe\cn\android\collegeinfo\CollegeInfoFragment.java

```java
1  public class CollegeInfoFragment extends Fragment {
2      private TextView infoView;              //显示学院信息文本
3      public View onCreateView(LayoutInflater inflater,ViewGroup container,
4              Bundle savedInstanceState){
5          //将布局文件转换成View对象
6          View collegeView=inflater.inflate(R.layout.college,container,false);
7          infoView= (TextView)collegeView.findViewById(R.id.infoView);
8          //学院信息比较多,在此保存在一个TXT文件中,通过I/O流进行读取
9          InputStream inputStream=getResources().openRawResource
                (R.raw.college_info);
10         infoView.setText(getStringFromInputStream(inputStream));
11         return collegeView;
12     }
13     //从输入流中读取字符串
14     public String getStringFromInputStream(InputStream inputStream){
15         byte[] buffer=new byte[1024];
16         int hasRead=0;                      //记录读取的字节个数
17         StringBuilder result=new StringBuilder("");
18         try {
```

```
19            while((hasRead=inputStream.read(buffer))!=-1){
20                //根据读取的字节构建字符串,并添加到已有字符串后面
21                result.append(new String(buffer,0,hasRead,"GBK"));
22            }
23        } catch(Exception ex){
24            ex.printStackTrace();
25        }
26        return result.toString();
27    }
28 }
```

现任领导页面对应的布局文件以及主程序代码如下。

页面布局文件:13\CollegeInfo\res\layout\leader.xml

```
1  <LinearLayout xmlns:android="http://schemas.android.com/apk/res/android"
2      android:layout_width="match_parent"
3      android:layout_height="match_parent"
4      android:background="#aabbcc"
5      android:orientation="vertical">
6      <TextView
7          android:layout_width="match_parent"
8          android:layout_height="wrap_content"
9          android:gravity="center"
10         android:textSize="24sp"
11         android:background="#77ccbbaa"
12         android:padding="10dp"
13         android:text="@string/leaderTitle"/>
14     <ListView
15         android:id="@+id/mListView"
16         android:layout_width="match_parent"
17         android:layout_height="wrap_content"
18         android:divider="#666666"
19         android:dividerHeight="2dp"/>
20 </LinearLayout>
```

列表中单项的布局文件:13\CollegeInfo\res\layout\item.xml

```
1  <LinearLayout xmlns:android="http://schemas.android.com/apk/res/android"
2      xmlns:tools="http://schemas.android.com/tools"
3      android:layout_width="match_parent"
4      android:layout_height="match_parent"
5      android:orientation="horizontal" >
6      <ImageView
7          android:id="@+id/image"
```

```
 8            android:layout_width="60dp"
 9            android:layout_height="72dp"
10            android:layout_margin="5dp"
11            android:contentDescription="@string/imgInfo"/>
12      <LinearLayout
13            android:layout_width="wrap_content"
14            android:layout_height="match_parent"
15            android:gravity="center_vertical"
16            android:orientation="vertical" >
17          <TextView
18              android:id="@+id/name"
19              android:layout_width="wrap_content"
20              android:layout_height="wrap_content"
21              android:textColor="#000000"
22              android:textSize="20sp"/>
23          <TextView
24              android:id="@+id/job"
25              android:layout_width="wrap_content"
26              android:layout_height="wrap_content"
27              android:layout_marginTop="10dp"
28              android:gravity="left"
29              android:textColor="#0000ee"
30              android:textSize="16sp"/>
31      </LinearLayout>
32  </LinearLayout>
```

页面主程序：13\CollegeInfo\src\iet\jxufe\cn\android\collegeinfo\LeaderFragment.java

```
 1  public class LeaderFragment extends Fragment {
 2      private ListView mListView;            //列表控件
 3      private String[] names=new String[]{"关爱浩","李新海","黄茂军","白耀
                          辉","邓庆山","彭敏"};
 4      private int[] imgIds=new int[] { R.drawable.guanaihao,
 5          R.drawable.lixinhai,R.drawable.huangmaojun,R.drawable.baiyaohui,
 6          R.drawable.dengqingshan,R.drawable.pengmin};
 7      private String[] jobs=new String[] { "院长","书记","教学副院长",
                          "科研副院长","学科建设副院长","学院副书记"};
 8      private List<Map<String,Object>>list=new ArrayList<Map<String,
            Object>>();
 9      public View onCreateView(LayoutInflater inflater,ViewGroup container,
10              Bundle savedInstanceState){
11          View leaderView=inflater.inflate(R.layout.leader,container,false);
12          mListView= (ListView)leaderView.findViewById(R.id.mListView);
13          init();
```

```
14        SimpleAdapter adapter=new SimpleAdapter(getActivity(),list,
             R.layout.item,new String[] {"name","img","job"},
15           new int[] {R.id.name,R.id.image,R.id.job});
16
17        mListView.setAdapter(adapter);
18        return leaderView;
19    }
20    public void init(){
21        for (int i=0; i<names.length; i++){
22            Map<String,Object>item=new HashMap<String,Object>();
23            item.put("name","姓名:"+names[i]);
24            item.put("img",imgIds[i]);
25            item.put("job","职务:"+jobs[i]);
26            list.add(item);
27        }
28    }
29 }
```

院属部门页面对应的布局文件以及主程序代码如下。

院属部门页面布局文件：13\CollegeInfo\res\layout\department.xml

```
1  <LinearLayout xmlns:android="http://schemas.android.com/apk/res/android"
2     android:layout_width="match_parent"
3     android:layout_height="match_parent"
4     android:background="#aabbcc"
5     android:orientation="vertical">
6     <TextView
7        android:layout_width="match_parent"
8        android:layout_height="wrap_content"
9        android:gravity="center"
10       android:textSize="24sp"
11       android:background="#77ccbbaa"
12       android:padding="10dp"
13       android:text="@string/departmentTitle"/>
14    <ListView
15       android:id="@+id/departmentView"
16       android:textSize="16sp"
17       android:textColor="#aa0000aa"
18       android:layout_width="match_parent"
19       android:layout_height="wrap_content"
20       android:paddingLeft="10dp"/>
21 </LinearLayout>
```

院属部门页面主程序：13\CollegeInfo\src\iet\jxufe\cn\android\collegeinfo\DepartmentFragment.java

```java
1   public class DepartmentFragment extends Fragment {
2       private ListView mListView;         //列表控件
3       private String[] names=new String[]{"电子工程系","网络工程系",
                    "软件工程系","通信工程系","培训中心"};
4       public View onCreateView(LayoutInflater inflater,ViewGroup container,
5               Bundle savedInstanceState){
6           View leaderView=inflater.inflate(R.layout.department,container,false);
7           mListView=(ListView)leaderView.findViewById(R.id.departmentView);
8           ArrayAdapter<String>adapter=new ArrayAdapter<String>(getActivity(),
                android.R.layout.simple_list_item_1,names);
9           mListView.setAdapter(adapter);
10          return leaderView;
11      }
12  }
```

主程序：13\CollegeInfo\src\iet\jxufe\cn\android\collegeinfo\MainActivity.java

```java
1   public class MainActivity extends Activity{
2       private TabHost mTabHost;                //选项卡控件
3       private int[] icons=new int[]{R.drawable.college,R.drawable.leader,
                R.drawable.department};       //选项卡图标
4       private String[] tags=new String[]{"college","leader","department"};
                                                 //选项卡标记
5       private String[] titles=new String[]{"学院简介","现任领导","院属部门"};
                                                 //选项卡标题
6       protected void onCreate(Bundle savedInstanceState){
7           super.onCreate(savedInstanceState);
8           requestWindowFeature(Window.FEATURE_NO_TITLE);  //不显示标题栏
9           setContentView(R.layout.activity_main);
10          mTabHost=(TabHost)findViewById(R.id.mTabHost);  //获取选项卡控件
11          mTabHost.setup();
12          for(int i=0;i<titles.length;i++){            //循环添加选项卡
13              TabSpec tabSpec=mTabHost.newTabSpec(tags[i]);
                //创建一个选项,并制定其标记
14              View view=getLayoutInflater().inflate(R.layout.tab, null);
                //将布局文件转换为 View 对象
15              TextView titleView=(TextView)view.findViewById(R.id.title);
16              ImageView iconView=(ImageView)view.findViewById(R.id.icon);
17              titleView.setText(titles[i]);        //设置标题
18              iconView.setImageResource(icons[i]); //设置图标
19              tabSpec.setIndicator(view);          //为选项设置标题和图标
20              tabSpec.setContent(R.id.realcontent);//为选项设置内容
21              mTabHost.addTab(tabSpec);            //将选项添加到选项卡中
```

```
22          }
23          mTabHost.setOnTabChangedListener(new MyTabChangedListener());
            //添加选项改变事件处理
24          mTabHost.setCurrentTabByTag("leader");        //设置初始显示的选项页
25      }
26      private class MyTabChangedListener implements OnTabChangeListener{
            //自定义选项改变事件监听器
27          public void onTabChanged(String tabTag){
28              FragmentTransaction fragmentTransaction=getFragmentManager()
                        .beginTransaction();                //开始事务
29              //判断单击的是哪个选项卡
30              if(tabTag.equalsIgnoreCase("leader")){
31                  fragmentTransaction.replace(R.id.realcontent,
                        new LeaderFragment(), "leader");
32              }else if(tabTag.equalsIgnoreCase("college")){
33                  fragmentTransaction.replace(R.id.realcontent,
                        new CollegeInfoFragment(), "college");
34              }else if(tabTag.equalsIgnoreCase("department")){
35                  fragmentTransaction.replace(R.id.realcontent,
                        new DepartmentFragment(), "department");
36              }
37              fragmentTransaction.commit();               //提交事务
38          }
39      }
40  }
```

13.3 代码分析

13.3.1 TabHost 介绍

TabHost 控件是 Android 应用中比较实用的控件，常用于页面的导航、切换，可以很方便地在一个 Activity 中实现多个页面的切换。它主要包含两个部分，即托盘（TabWidget）和显示具体页面信息的 FrameLayout（TabContent）。TabWidget 主要用于显示不同的选项，如本例中包含学院介绍、现任领导、院属信息 3 个选项，TabWidget 根据用户需求既可以放置在页面的顶部，也可以放置在页面的底部。

TabHost 中的每个选项用 TabSpec 来表示，可以为它设置标记、标题、图标以及该选项对应的内容等信息，单击某个 TabSpec 后，会在 TabContent 中显示其对应的信息。TabSpec 是 TabHost 类的一个内部类，本身不向外提供公有的构造方法，因此不能通过 new 关键字来实例化，需要调用 TabHost 的 newSpec()方法创建 TabSpec，在创建一个选项之后，对其进行简单的设置，最后将其添加到 TabHost 中，具体代码见 MainActivity 的第 13~21 行。

与 ListView 类似，使用 TabHost 有两种方式，一种是在布局文件中放置 TabHost 控件，然后根据 ID 找到 TabHost，再进行相关的操作；另一种是从系统提供的 TabActivity

继承而来,此时页面中会自动包含一个TabHost,通过TabActivity的getTabHost()方法可获取TabHost,然后进行相关的操作。使用继承TabActivity的方式能够非常方便地实现TabHost效果,而通过在布局文件中自定义TabHost相对来说比较麻烦。TabActivity在Android API Level 13中已经被废弃了,Android官方文档推荐使用Fragment。本案例介绍如何在布局中自定义TabHost来实现切换功能。

在布局文件中定义TabHost之后,还必须在其内部添加TabWidget和FrameLayout两个控件,为它们添加ID属性,并且它们的ID属性值是固定的,分别为系统定义的@android:id/tabs和@android:id/tabcontent,而TabHost控件的ID没有任何要求,这是因为通过findViewById()方法获取的TabHost控件在加载选项(TabSpec)之前必须调用setUp()方法,在该方法内部有以下语句。

```
1  mTabWidget=(TabWidget)findViewById(com.android.internal.R.id.tabs);
2  if(mTabWidget==null){
3      throw new RuntimeException("Your TabHost must have a TabWidget whose id
                                    attribute is 'android.R.id.tabs'");
4  }
5  …
6  mTabContent=(FrameLayout)findViewById(com.android.internal.R.id.tabcontent);
7  if(mTabContent==null){
8      throw new RuntimeException("Your TabHost must have a FrameLayout whose id
                                    attribute is 'android.R.id.tabcontent'");
9  }
```

在该方法内部会根据com.android.internal.R.id.tabs来查找TabWidget控件,如果未找到,该控件则抛出异常提示用户:TabHost中一定要包含一个ID为android.R.id.tabs的TabWidget控件。根据com.android.internal.R.id.tabcontent来查找FrameLayout控件,如果未找到该控件,同样会抛出异常提示用户:TabHost中一定要包含一个ID为android.R.id.tabcontent的FrameLayout控件。TabWidget与FrameLayout的位置则可根据需求进行设定,在此设置TabWidget处于底部,FrameLayout填充剩余的所有空间。代码见activity_main.xml。

选中TabHost中的某个选项之后,需要动态地改变页面的显示,在此为TabHost添加页面变化事件监听器,一旦监听到页面变化,则改变FrameLayout中的内容,每个页面中的该部分内容都是通过Fragment来实现的。

13.3.2 Fragment介绍

Fragment是Android 3.0引入的新API,英文意思是片段,用户可以把它理解为Activity中的片段或者子模块。Fragment拥有自己的生命周期,也可以接受自己的输入事件。但Fragment必须被嵌入到Activity中使用,Fragment的生命周期会受它所在的Activity的生命周期的控制。例如,当Activity暂停时,该Activity内的所有Fragment都会暂停,当Activity被销毁时,该Activity内的所有Fragment都会被销毁。而当Activity处于运行状态时,我们可以独立地操作每一个Fragment,例如添加、删除等。关于 Fragment，主要有以下特点。

（1）在一个 Activity 中可以同时包含多个 Fragment；反过来，一个 Fragment 也可以被多个 Activity 复用。

（2）Fragment 总是作为 Activity 界面组成的一部分。在 Fragment 中，可通过 getActivity() 方法获取它所在的 Activity；在 Activity 中，可以通过 getFragmentManager() 方法得到 Fragment 管理器，然后调用它的 findFragmentById() 或者 findFragmentByTag() 方法获取 Fragment。

（3）Fragment 拥有自己的生命周期，也可以响应自己的输入事件，但它的生命周期直接受它所属的 Activity 的生命周期控制。

（4）只有当 Activity 处于活动状态时，才可以调用 FragmentTransaction 的 add()、remove()、replace() 方法动态地添加、删除、替换 Fragment。

与 Activity 类似，创建自定义的 Fragment 必须继承系统提供的 Fragment 基类或者它的子类，然后可根据需要实现它的一些方法。Fragment 中的回调方法与 Activity 的回调方法非常类似，主要包含 onAttach()、onCreate()、onCreateView()、onActivityCreated()、onStart()、onResume()、onPause()、onStop()、onDestroyView()、onDestroy()、onDetach()等。为了控制 Fragment 的显示，通常需要重写 onCreateView() 方法，该方法返回的 View 将作为该 Fragment 显示的 View 控件，当 Fragment 绘制界面时会回调该方法。在本案例中的 3 个 Fragment 都比较简单，仅仅重写了 onCreateView() 方法。

Fragment 在创建完成后还需要嵌入到 Activity 中，将 Fragment 添加到 Activity 中有以下两种方式：

（1）在布局文件中使用＜fragment.../＞标签添加 Fragment，通过该标签的 android：name 属性指定 Fragment 的实现类，属性值为完整的包名＋类名。

（2）在 Java 代码中，通过 getFragmentManager() 方法获取 FragmentManager 对象，然后调用其 beginTransaction() 方法开启事务，得到 FragmentTransaction 对象，再调用该对象的 add() 方法添加 Fragment，最后调用它的 commit() 方法提交事务。

在本案例中，由于需要动态地改变 Fragment，因此采用第二种方式，代码见第 **38**～**51** 行。

13.3.3 根据状态改变图片

在 Android 应用中，为了区分用户的操作，通常会根据状态来改变控件的背景或图片。实现该效果通常有两种方式，一是为控件添加相应的事件处理，然后在对应的方法中通过代码来改变其背景或者图片；二是定义一种特殊的 XML 图片，该图片会根据控件的状态显示相应的图片。其中，第一种方式相对来说比较麻烦，并且复用性不强，如果有多个这样的控件则需要单独为每个控件添加事件处理，代码较冗长，而第二种方式只需要定义一个 XML 图片，当控件需要使用时，直接引用即可。本案例中采用第二种方式。

在 Android 中，定义根据状态改变显示的 XML 图片文件对应的根标签是＜selector.../＞，该元素可以包含多个＜item.../＞元素，其中每个＜item.../＞元素表示一种状态，通过 item 元素可设置该状态对应的图片，item 元素主要包含以下两个属性。

（1）android：state_xxx：指定一个特定状态。

（2）android：drawable：指定该状态对应的图片。

item 元素中支持的状态主要如下。
- android:state_active：表示是否处于激活状态。
- android:state_checkable：表示是否处于可勾选状态。
- android:state_checked：表示是否处于已勾选状态。
- android:state_enabled：表示是否处于可用状态。
- android:state_first：表示是否处于开始状态。
- android:state_focused：表示是否处于已得到焦点状态。
- android:state_last：表示是否处于结束状态。
- android:state_middle：表示是否处于中间状态。
- android:state_pressed：表示是否处于已被按下状态。
- android:state_selected：表示是否处于已被选中状态。
- android:state_window_focused：表示窗口是否已得到焦点状态。

和其他 XML 标签一样，<selector.../>标签也有对应的 Java 类，该类的类名为 StateListDrawable，该类中提供了一个 addState(int[] stateSet，Drawable drawable)方法，该方法的功能类似于<item...>标签，用于指定某一或某些状态下对应的图片。

13.4 知识扩展

13.4.1 Fragment 与 Activity 交互

上面的例子只是简单地介绍了 Fragment 的使用，通过 Activity 来动态改变需要显示的 Fragment，并未涉及 Fragment 本身的事件处理。下面在之前的基础上为院属部门页面的列表项添加事件处理，选中某一项后，可以显示该项的详细介绍，单击"返回"按钮后，又可以回到院属部门列表页面，程序运行效果如图 13-4 和图 13-5 所示。

图 13-4　软件工程系介绍页面

图 13-5　网络工程系介绍页面

更改 DepartmentFragment.java 文件，为列表项添加事件处理，由于每一个部门的介绍文字都比较多，不宜直接写在程序代码中，在此将其存放在 TXT 文件中，并将该文件存放在 res/raw 文件夹下。DepartmentFragment 代码如下。

院属部门页面主程序：13\CollegeInfo\src\iet\jxufe\cn\android\collegeinfo\DepartmentFragment.java

```
1   public class DepartmentFragment extends Fragment {
2       private int[] infos=new int[] { R.raw.dianzi, R.raw.wangluo,
3           R.raw.ruanjian, R.raw.tongxin, R.raw.peixun };   //存放部门详细信息的文件
4       private ListView mListView;                          //列表控件
5       private String[] names=new String[] { "电子工程系", "网络工程系",
            "软件工程系", "通信工程系", "培训中心" };
6       public View onCreateView(LayoutInflater inflater, ViewGroup container,
7               Bundle savedInstanceState) {
8           View leaderView=inflater.inflate(R.layout.department,
                container, false);
9           mListView= (ListView) leaderView.findViewById(R.id.departmentView);
10          ArrayAdapter<String>adapter=new ArrayAdapter<String>(getActivity(),
11              android.R.layout.simple_list_item_1, names);
12          mListView.setAdapter(adapter);
13          mListView.setOnItemClickListener(new MyItemClickListener());
14          return leaderView;
15      }
16      private class MyItemClickListener implements OnItemClickListener {
17          public void onItemClick(AdapterView<?>parent, View view, int position,
18                  long id) {
19              Bundle bundle=new Bundle();         //创建 Bundle 对象,用于传递数据
20              bundle.putInt("detailId", infos[position]);   //传递标题
21              bundle.putString("title", names[position]);   //传递内容文件 ID
22              DepartmentInfoFragment depInfoFragment=new
                    DepartmentInfoFragment();
                //创建 Fragment 对象
23              depInfoFragment.setArguments(bundle);          //设置参数
24              getActivity().getFragmentManager().beginTransaction()
25                  .replace(R.id.realcontent, depInfoFragment).commit();
                //用新的 Fragment 替换当前的 Fragment
26          }
27      }
28  }
```

显示部门详细信息的布局文件：13\CollegeInfo\res\layout\department_info.xml

```
1   <LinearLayout xmlns:android="http://schemas.android.com/apk/res/android"
2       android:layout_width="match_parent"
```

```
3          android:layout_height="match_parent"
4          android:background="#aabbcc"
5          android:orientation="vertical" >
6          <RelativeLayout
7              android:layout_width="match_parent"
8              android:layout_height="wrap_content"
9              android:background="#77ccbbaa">
10             <TextView
11                 android:id="@+id/infoTitle"
12                 android:layout_width="match_parent"
13                 android:layout_height="wrap_content"
14                 android:gravity="center"
15                 android:padding="10dp"
16                 android:textSize="24sp"/>
17             <ImageButton
18                 android:id="@+id/goBack"
19                 android:src="@drawable/goback"
20                 android:layout_alignParentRight="true"
21                 android:layout_centerVertical="true"
22                 android:layout_width="wrap_content"
23                 android:layout_height="wrap_content"
24                 android:background="#00000000"
25                 android:layout_marginRight="20dp"
26                 android:contentDescription="@string/imgInfo"/>
27         </RelativeLayout>
28         <ScrollView
29             android:layout_width="match_parent"
30             android:layout_height="wrap_content" >
31             <TextView
32                 android:id="@+id/detailInfo"
33                 android:layout_width="match_parent"
34                 android:layout_height="wrap_content"
35                 android:padding="5dp"
36                 android:textColor="#004400"
37                 android:textSize="16sp"/>
38         </ScrollView>
39  </LinearLayout>
```

详细信息页面程序：13\CollegeInfo\src\iet\jxufe\cn\android\collegeinfo\DepartmentInfoFragment.java

```
1   public class DepartmentInfoFragment extends Fragment {
2       private TextView infoTitle,detailInfo;         //显示标题和内容的文本控件
3       private ImageButton goBack;                    //"返回"按钮
4       private String titleString="电子工程";          //保存标题值
```

```java
5       private int detailId=R.raw.dianzi;              //保存内容对应文件的ID
6       public void onCreate(Bundle savedInstanceState){
7           super.onCreate(savedInstanceState);
8           Bundle bundle=getArguments();               //获取传递过来的参数
9           if(bundle!=null){
10              titleString=bundle.getString("title","电子工程");   //从参数中获取标题
11              detailId=bundle.getInt("detailId",R.raw.dianzi);
                //从参数中获取内容文件ID
12          }
13      }
14      public View onCreateView(LayoutInflater inflater,ViewGroup container,
15              Bundle savedInstanceState){
16          View detailView=inflater.inflate(R.layout.department_info,
                container,false);
17          detailInfo=(TextView)detailView.findViewById(R.id.detailInfo);
18          infoTitle=(TextView)detailView.findViewById(R.id.infoTitle);
19          goBack=(ImageButton)detailView.findViewById(R.id.goBack);
20          infoTitle.setText(titleString);
21          InputStream inputStream=getResources().openRawResource(detailId);
22          detailInfo.setText(getStringFromInputStream(inputStream));
23          goBack.setOnClickListener(new OnClickListener(){
24              public void onClick(View v){
25                  getActivity().getFragmentManager().beginTransaction()
26                      .replace(R.id.realcontent,new DepartmentFragment())
27                      .commit();
28              }
29          });
30          return detailView;
31      }
32      //从输入流中读取字符串
33      public String getStringFromInputStream(InputStream inputStream){
34          byte[] buffer=new byte[1080];
35          int hasRead=0;                              //记录读取的字节个数
36          StringBuilder result=new StringBuilder("");
37          try {
38              while((hasRead=inputStream.read(buffer))!=-1){
39                  //根据读取的字节构建字符串,并添加到已有字符串后面
40                  result.append(new String(buffer,0,hasRead,"GBK"));
41              }
42          } catch(Exception ex){
43              ex.printStackTrace();
44          }
45          return result.toString();
46      }
```

47   }

当需要向 Fragment 中传递参数时，可创建 Bundle 数据包，然后调用 Fragment 的 setArgument(Bundle bundle)方法将 Bundle 数据包传递给 Fragment。接下来在 Fragment 中可通过 getArgument()方法获取到该数据包，再进行相关操作。

13.4.2　ActionBar 实现页面切换效果

除了可以使用 TabHost 实现导航、切换效果以外，在 Android 3.0 之后的版本中新增了 ActionBar，也可以很方便地实现该效果，并且使用相对比较简单。使用 ActionBar 实现上例的效果如图 13-6 至图 13-8 所示。

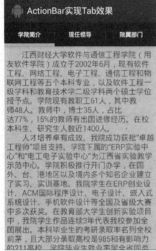
图 13-6　学院简介页面

图 13-7　现任领导介绍页面

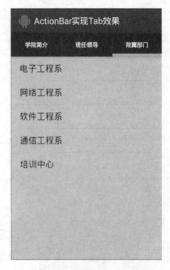
图 13-8　院属部门页面

每个页面内容的显示和上面相同，仍然是使用 Fragment，和上例的差异在于主界面和主程序，具体代码如下。

主界面布局文件：13\ActionBarTab\res\layout\activity_main.xml

```
1   <RelativeLayout xmlns:android="http://schemas.android.com/apk/res/android"
2       android:id="@+id/container"
3       android:layout_width="match_parent"
4       android:layout_height="match_parent"
5       android:background="#aabbcc">
6   </RelativeLayout>
```

主程序代码：13\ActionBarTab\src\iet\jxufe\cn\android\actionbartab\MainActivity.java

```
1   public class MainActivity extends Activity {
2       private ActionBar mActionBar;
3       private String[] titles=new String[]{"学院简介","现任领导","院属部门"};
```

```
                //选项卡标题
4       protected void onCreate(Bundle savedInstanceState) {
5           super.onCreate(savedInstanceState);
6           setContentView(R.layout.activity_main);
7           mActionBar=getActionBar();              //获取 ActionBar
8           mActionBar.setNavigationMode(ActionBar.NAVIGATION_MODE_TABS);
            //设置 ActionBar 的导航模式
9           MyTabListener mTabListener=new MyTabListener();
            //创建自定义的页面监听器
10          for(int i=0;i<titles.length;i++){       //循环向 ActionBar 中添加选项
11              Tab tab=mActionBar.newTab();        //创建一个选项页
12              tab.setText(titles[i]);             //设置选项页的标题
13              tab.setTabListener(mTabListener);   //为选项页添加页面监听器
14              mActionBar.addTab(tab);             //将选项页添加到 ActionBar 中
15          }
16      }
17      private class MyTabListener implements TabListener{    //页面监听器
18          public void onTabSelected(Tab tab, FragmentTransaction ft) {
                //选项页被选中的事件处理
19              String tabText=tab.getText().toString();  //获取选中的选项页的标题
20              FragmentTransaction fragmentTransaction=getFragmentManager().
                    beginTransaction();             //开启事务
21              if(tabText.equalsIgnoreCase("现任领导")){
                    //判断选中的是哪一个页面,然后用新的 Fragment 替换原有内容
22                  fragmentTransaction.replace(R.id.container, new
                        LeaderFragment());
23              }else if(tabText.equalsIgnoreCase("学院简介")){
24                  fragmentTransaction.replace(R.id.container,new
                        CollegeInfoFragment());
25              }else if(tabText.equalsIgnoreCase("院属部门")){
26                  fragmentTransaction.replace(R.id.container,new
                        DepartmentFragment());
27              }
28              fragmentTransaction.commit();       //提交事务
29          }
30          public void onTabUnselected(Tab tab, FragmentTransaction ft) {}
            //选项页未被选中的事件处理
31          public void onTabReselected(Tab tab, FragmentTransaction ft) {}
            //选项页再次被选中的事件处理
32      }
33  }
```

通过上述程序可知,虽然通过 ActionBar 实现 Tab 效果较为简单,但也存在缺陷。ActionBar 是系统固定的,只能处于屏幕的顶部而不能像 TabHost 那样灵活摆放,默认情

况下不能去除应用的图标和标题，空间相对较少，每个选项的内容不能任意定义，在 ActionBar 中 Tab 类虽然提供了 setIcon() 方法，用于指定选项的图标，但是标题和图标只能水平摆放，而在 TabHost 的 TabSpec 中可以传递一个 View 作为选项，这就非常灵活。总之，二者各有利弊，用户需根据具体情况选择使用，当需求非常简单时，建议使用 ActionBar，当需求较为复杂、灵活性、扩展性要求较高时，建议使用自定义 TabHost。

注意：ActionBar 是 Android 3.0 之后才提供的新特性，因此，需要在 AndroidManifest 文件中将应用程序的最低版本设置为 11。

13.5 思考与练习

（1）改变 TabHost 实现页面切换效果部分，使得选项在顶部显示。

（2）完善 ActionBar 实现页面切换效果部分，使得选中院属部门列表中的某一项后可以显示该部门的详细介绍信息。

第14章 省市二级列表——ExpandableListView 的应用

14.1 案例概述

本案例主要实现省市二级列表的效果,介绍 ExpandableListView 控件的使用。其中,省份是一个列表,展开列表中的某一项后会显示一个城市的列表。本案例的关键就是如何将城市列表与所在的省份关联起来。与 ListView 类似,ExpandableListView 本身并不能与数据源关联,需要借助相应的 Adapter 协助,可以自定义也可以使用系统提供的 Adapter 实现类。由于本案例中每一项的数据相对来说比较简单,仅仅是显示文本的 TextView,因此使用系统提供的 Adapter——SimpleExpandableListAdapter。本案例的程序运行效果如图 14-1 和图 14-2 所示。

图 14-1 程序运行效果图 1

图 14-2 程序运行效果图 2

14.2 关键代码

主界面布局文件:14\ProvinceAndCityList\res\layout\activity_main.xml

```
1  <LinearLayout xmlns:android="http://schemas.android.com/apk/res/android"
2      xmlns:tools="http://schemas.android.com/tools"
```

```
3       android:layout_width="match_parent"
4       android:layout_height="match_parent"
5       android:background="#aabbcc">
6       <ExpandableListView
7           android:id="@+id/mExpandableListView"
8           android:layout_width="match_parent"
9           android:layout_height="wrap_content"/>
10  </LinearLayout>
```

显示省份的布局文件：14\ProvinceAndCityList\res\layout\province.xml

```
1   <LinearLayout xmlns:android="http://schemas.android.com/apk/res/android"
2       xmlns:tools="http://schemas.android.com/tools"
3       android:layout_width="match_parent"
4       android:layout_height="match_parent"
5       android:gravity="center_vertical">
6       <TextView
7           android:id="@+id/group"
8           android:paddingLeft="40dp"
9           android:layout_width="match_parent"
10          android:layout_height="wrap_content"
11          android:textColor="#ff0000"
12          android:paddingTop="10dp"
13          android:paddingBottom="10dp"
14          android:textSize="20sp" />
15  </LinearLayout>
```

显示城市的布局文件：14\ProvinceAndCityList\res\layout\city.xml

```
1   <LinearLayout xmlns:android="http://schemas.android.com/apk/res/android"
2       xmlns:tools="http://schemas.android.com/tools"
3       android:layout_width="match_parent"
4       android:layout_height="match_parent"
5       android:gravity="center_vertical">
6       <TextView
7           android:id="@+id/child"
8           android:layout_width="match_parent"
9           android:layout_height="wrap_content"
10          android:textColor="#0000ff"
11          android:paddingLeft="60dp"
12          android:paddingTop="5dp"
13          android:paddingBottom="5dp"
14          android:textSize="16sp" />
15  </LinearLayout>
```

主程序文件：14\ProvinceAndCityList\src\iet\jxufe\cn\android\provinceandcitylist\MainActivity.java

```java
1   public class MainActivity extends Activity{
2       private ExpandableListView mExpandableListView;         //扩展下拉列表
3       private String[] provinces=new String[]{"江西","江苏","浙江"}; //省份信息
4       private String[][] cities=new String[][]{{"南昌","九江","赣州",
            "吉安"},{"南京","苏州","南通"},{"杭州","金华"}};         //城市信息
5       private List<Map<String,String>>provinceItems=new ArrayList<Map<String,
            String>>();                                         //保存所有省份信息的集合
6       private List<List<Map<String,String>>>cityItems=new ArrayList
            <List<Map<String,String>>>();                       //保存所有城市信息的集合
7       protected void onCreate(Bundle savedInstanceState){
8           super.onCreate(savedInstanceState);
9           this.setContentView(R.layout.activity_main);
10          mExpandableListView=(ExpandableListView)findViewById
                (R.id.mExpandableListView);
11          init();                                             //执行初始化操作
12          SimpleExpandableListAdapter adapter=new SimpleExpandableListAdapter(
                this,provinceItems,R.layout.province,new String[]{"group"},
                new int[]{R.id.province},cityItems,R.layout.city,
                new String[]{"child"},new int[]{R.id.city});
            //使用系统提供的Adapter
13          mExpandableListView.setAdapter(adapter);            //关联数据
14      }
15      public void init(){                                     //执行初始化操作
16          for(int i=0;i<provinces.length;i++){
            //循环遍历省份,将省份与相应的城市关联
17              Map<String,String>provinceItem=new HashMap<String,String>();
18              provinceItem.put("group",provinces[i]);
19              provinceItems.add(provinceItem);
20              //该省份所包含城市的集合
21              List<Map<String,String>>cityList=new ArrayList<Map<String,
                    String>>();
22              for(int j=0;j<cities[i].length;j++){
23                  Map<String,String>cityItem=new HashMap<String,String>();
24                  cityItem.put("child",cities[i][j]);
25                  cityList.add(cityItem);
26              }
27              cityItems.add(cityList);                        //所有省份所包含城市的集合
28          }
29      }
30  }
```

14.3 代码分析

本案例讲解的是一种特殊的列表,该列表中的每一项展开后又是一个列表,在实际应用中非常常见,例如每个省份下面又有很多城市、每类产品下面又有很多子产品、每章下面又有很多节等。

和 ListView 类似,用户可以把列表中的每一项信息保存在一个集合中,将省份信息放在一个集合中,城市信息放在一个集合中,但城市之间又存在一定的关系,即是否属于同一个省份,需要对其进行分类,所以城市集合是一种比较特殊的集合,该集合中的一个元素表示属于同一个省的城市,也是一个集合,因此城市集合的声明为"List<List<Map<String,String>>>cityItems"。

有了这些数据源之后还需要将这些数据源与列表控件关联起来,在此是通过系统提供的 SimpleExpandableListAdapter 来实现的。在创建该对象时,需要传递 9 个参数,该类的构造方法为"SimpleExpandableListAdapter(Context context, List<? extends Map<String, ?>> groupData, int groupLayout, String[] groupFrom, int[] groupTo, List<? extends List<? extends Map<String, ?>>> childData, int childLayout, String[] childFrom, int[] childTo)",其 9 个参数的含义如下。

- context:表示上下文对象,通常传递当前的 Activity 对象。
- groupData:表示一级列表数据的集合,在此为省份的集合 provinceItems。
- groupLayout:表示一级列表数据项所对应的布局文件,在此为 province.xml 文件。
- groupFrom:表示获取一级列表数据集合中数据的关键字所组成的数组,即 provinceItem 对象中的相关 key,根据这些 key 可以获取指定的数据值。
- groupTo:表示显示一级列表中数据的控件 ID 所组成的数组,在获取一级列表项中的数据后,需要将这些数据显示在相应的控件之上,控件通过 ID 唯一标识,groupTo 与 groupFrom 参数之间存在一一对应关系,并且 groupTo 中的 ID 必须在 groupLayout 布局文件中。
- childData:表示所有二级列表项的集合,在此为城市的集合 cityItems。
- childLayout:表示二级列表项所对应的布局文件,在此为 city.xml 文件。
- childFrom:表示获取二级列表集合中数据的关键字所组成的数组,即 cityItem 对象中的相关 key,根据这些 key 可以获取指定的数据值。
- childTo:表示显示二级列表中数据的控件 ID 所组成的数据,在获取二级列表项中的数据后,将这些数据显示在对应的控件之上,childTo 与 childFrom 存在一一对应关系,并且 childTo 中的 ID 必须在 childLayout 布局文件中。

通过使用系统为我们提供的 SimpleExpandableListAdapter 可以很方便地实现扩展下拉列表的功能,但是该功能非常有限,列表项的内容只能是文本,如果想显示比较复杂的列表项,如列表项中包含图片,则需要使用自定义的 Adapter。

14.4 知识扩展

与 ListView 类似，用户除了可以使用系统提供的一些常见的 Adapter 之外，也可以通过继承系统提供的 Adapter 基类来自定义 Adapter。对于扩展下拉列表来说，Adapter 对应的基类为 BaseExpandableListAdapter()，在使用自定义的 Adapter 时，需要重写该类中的相关方法，虽然代码较多，但是更为灵活，不受限制。下面通过自定义 Adapter 来实现较为复杂的二级列表，在每个城市的左边添加一张图片，程序运行效果如图 14-3 和图 14-4 所示。

图 14-3　省市二级列表

图 14-4　省份展开效果图

主界面布局代码和显示省份信息的布局文件代码同上，在此不再列出。

显示城市的布局文件：14\ProvinceAndCityList\res\layout\city.xml

```
1   <LinearLayout xmlns:android="http://schemas.android.com/apk/res/android"
2       xmlns:tools="http://schemas.android.com/tools"
3       android:layout_width="match_parent"
4       android:layout_height="match_parent"
5       android:orientation="horizontal"
6       android:gravity="center_vertical">
7       <ImageView
8           android:id="@+id/cityImg"
9           android:layout_width="50dp"
10          android:layout_height="30dp"
11          android:layout_marginTop="10dp"
12          android:layout_marginRight="10dp"
13          android:layout_marginBottom="10dp"
```

```
14          android:layout_marginLeft="40dp"
15          android:contentDescription="@string/imgInfo"/>
16      <TextView
17          android:id="@+id/city"
18          android:layout_width="match_parent"
19          android:layout_height="wrap_content"
20          android:textColor="#0000ff"
21          android:paddingTop="5dp"
22          android:paddingBottom="5dp"
23          android:textSize="16sp" />
24  </LinearLayout>
```

主程序:14\ProvinceAndCityDIY\src\iet\jxufe\cn\android\provinceandcitydiy\MainActivity.java

```
1   public class MainActivity extends Activity {
2       private ExpandableListView mExpandableListView;        //扩展下拉列表
3       private String[] provinces=new String[] {"江西","江苏","浙江"}; //省份信息
4       private String[][] cities=new String[][] {{"南昌","九江","赣州","吉安"},
5           {"南京","苏州","南通"},{"杭州","金华"}};              //城市信息
6       private int[][] cityImgIds=new int[][] {
7               {R.drawable.nanchang,R.drawable.jiujiang,R.drawable.ganzhou,
8               R.drawable.jian},{R.drawable.nanjing,R.drawable.suzhou,
9               R.drawable.nantong},{R.drawable.hangzhou,R.drawable.jinhua}};
            //城市图片信息
10      protected void onCreate(Bundle savedInstanceState){
11          super.onCreate(savedInstanceState);
12          this.setContentView(R.layout.activity_main);
13          mExpandableListView=(ExpandableListView)findViewById
                (R.id.mExpandableListView);
14          mExpandableListView.setAdapter(new MyAdapter());
15      }
16      private class MyAdapter extends BaseExpandableListAdapter{
        //自定义 Adapter 类
17          public int getGroupCount(){            //获取一级列表中包含的项数
18              return provinces.length;
19          }
20          public int getChildrenCount(int groupPosition){ //获取指定项所包含的子项数
21              return cities[groupPosition].length;
22          }
23          public Object getGroup(int groupPosition){    //获取指定的一级列表项
24              return null;
25          }
26          public Object getChild(int groupPosition,int childPosition){
            //获取指定的二级列表项
```

```
27              return null;
28          }
29          public long getGroupId(int groupPosition){   //获取指定的一级列表项的ID
30              return 0;
31          }
32          public long getChildId(int groupPosition,int childPosition){
            //获取指定的二级列表项的ID
33              return 0;
34          }
35          public boolean hasStableIds(){              //返回是否包含稳定的ID
36              return false;
37          }
38          public View getGroupView(int groupPosition,boolean isExpanded,
39                  View convertView,ViewGroup parent){    //获取一级列表项显示的控件
40              View groupView=getLayoutInflater().inflate
                    (R.layout.province,null);
41              TextView provinceText=(TextView)groupView.findViewById
                    (R.id.province);
42              provinceText.setText(provinces[groupPosition]);
43              return groupView;
44          }
45          public View getChildView(int groupPosition,int childPosition,
46              boolean isLastChild,View convertView,ViewGroup parent){
                //获取二级列表项显示的控件
47              View childView=getLayoutInflater().inflate(R.layout.city,null);
48              TextView cityText=(TextView)childView.findViewById(R.id.city);
49              ImageView cityImg=(ImageView)childView.findViewById(R.id.cityImg);
50              cityText.setText(cities[groupPosition][childPosition]);
51              cityImg.setImageResource(cityImgIds[groupPosition]
                    [childPosition]);
52              return childView;
53          }
54          public boolean isChildSelectable(int groupPosition,
                int childPosition){
                //子项是否可选
55              return false;
56          }
57      }
58  }
```

在通过继承 BaseExpandableListAdapter 实现自定义 Adapter 时,需要重写该类中的10个抽象方法,其中最为关键的是 getGroupCount()、getChildrenCount()、getGroupView、getChildView 这4个方法,根据这4个方法就可以获取一共有多少组、每一组中又包含多少子项、每一组如何显示以及组中的每一项如何显示。其他方法根据需要有选择性地重

写,如果没有要求,可采用默认值。例如 isChildSelectable()方法表示子项是否可选择,如果需要为子项添加选择事件处理,则该方法必须返回为 true,否则无法执行选择事件处理。

14.5 思考与练习

本案例中省份与城市的相关信息都是通过程序临时指定的,尝试建立一个数据库,保存相关信息,然后通过分类查询获取相关信息,再将其显示在扩展列表中。

第15章

产品分类——自定义多级列表效果

15.1 案例概述

本案例主要实现自定义多级列表的效果,在实际应用中存在很多层次结构体系,例如物品的分类、人类的继承关系等,而在 Android 中只为用户提供了一级列表 ListView 和二级列表 ExpandableListView 两种控件,远远满足不了用户的需求,那么此时用户就要根据已有知识变通一下,设计出类似效果。本案例实质上使用的仍然是 ListView,只不过采用的 Adapter 是自定义的 Adapter,在这个 Adapter 中做了一些处理。在单击每一项时,判断该项是否还有子项,如果有,则判断该项是否已经展开,如果没有展开,则展开显示子项;如果已经展开,则关闭。本案例的程序运行效果如图 15-1 和图 15-2 所示。

图 15-1 程序运行效果图 1

图 15-2 程序运行效果图 2

15.2 关键代码

主界面布局文件:15\ProductCatagories\res\layout\activity_main.xml

```
1   <RelativeLayout xmlns:android="http://schemas.android.com/apk/res/android"
2       xmlns:tools="http://schemas.android.com/tools"
```

```
3       android:layout_width="match_parent"
4       android:layout_height="match_parent"
5       android:background="#aabbcc">
6       <ListView
7           android:id="@+id/mListView"
8           android:layout_width="match_parent"
9           android:layout_height="wrap_content"/>
10  </RelativeLayout>
```

列表中每一项的布局文件：15\ProductCatagories\res\layout\treeview_item.xml

```
1   <LinearLayout xmlns:android="http://schemas.android.com/apk/res/android"
2       android:layout_width="wrap_content"
3       android:layout_height="wrap_content"
4       android:gravity="center_vertical"
5       android:orientation="horizontal" >
6       <ImageView
7           android:id="@+id/icon"
8           android:layout_width="wrap_content"
9           android:layout_height="40dp"
10          android:contentDescription="@string/imgInfo"/>
11      <TextView
12          android:id="@+id/text"
13          android:textSize="20sp"
14          android:layout_width="wrap_content"
15          android:layout_height="wrap_content"
16          android:padding="5dp"
17          android:gravity="center"/>
18  </LinearLayout>
```

自定义列表项：15\ProductCatagories\src\iet\jxufe\cn\android\productcatagories\Element.java

```
1   public class Element {
2       private String text;              //文本信息内容
3       private int level;                //在层次结构中的级别,最顶级为 0
4       private int id;                   //该项元素的 ID
5       private int parendId;             //直接父元素的 ID,如果没有父元素,该值为-1
6       private boolean hasChildren;      //是否有子元素
7       private boolean isExpanded;       //该项是否展开
8       //定义两个常量,顶级元素 level 为 0,父元素 ID 为-1
9       public static final int NO_PARENT=-1;
10      public static final int TOP_LEVEL=0;
11      public Element(String text,int level,int id,int parendId,
12              boolean hasChildren,boolean isExpanded){        //构造方法
```

```
13          this.text=text;
14          this.level=level;
15          this.id=id;
16          this.parendId=parendId;
17          this.hasChildren=hasChildren;
18          this.isExpanded=isExpanded;
19      }
20      //生成相应的set方法和get方法,设置和获取相应的属性值
21      public boolean isExpanded(){
22          return isExpanded;
23      }
24      public void setExpanded(boolean isExpanded){
25          this.isExpanded=isExpanded;
26      }
27      public String getText(){
28          return text;
29      }
30      public void setText(String text){
31          this.text=text;
32      }
33      public int getLevel(){
34          return level;
35      }
36      public void setLevel(int level){
37          this.level=level;
38      }
39      public int getId(){
40          return id;
41      }
42      public void setId(int id){
43          this.id=id;
44      }
45      public int getParendId(){
46          return parendId;
47      }
48      public void setParendId(int parendId){
49          this.parendId=parendId;
50      }
51      public boolean isHasChildren(){
52          return hasChildren;
53      }
54      public void setHasChildren(boolean hasChildren){
55          this.hasChildren=hasChildren;
56      }
```

57 }

主程序代码:ProductCatagories\src\iet\jxufe\cn\android\productcatagories\MainActivity.java

```java
1   public class MainActivity extends Activity {
2       private ListView mListView;                         //列表控件
3       private MyAdapter myAdapter;                        //自定义 Adapter 对象
4       private ArrayList<Element>visibleElements;          //可见的元素集合
5       private ArrayList<Element>allElements;              //所有的元素集合
6       private int basePadding=20;                         //默认上层与下层左边距之间的差距
7       private int baseSize=2;                             //默认相邻级别字体相差 2px
8       private int baseHeight=8;                           //默认相邻级别图片高度相差 8dp
9       protected void onCreate(Bundle savedInstanceState){
10          super.onCreate(savedInstanceState);
11          setContentView(R.layout.activity_main);
12          init();                                         //初始化数据
13          mListView=(ListView)findViewById(R.id.mListView);
14          myAdapter=new MyAdapter();
15          mListView.setAdapter(myAdapter);
16          mListView.setOnItemClickListener(new MyItemClickListener());
17      }
18      private void init(){                                //执行初始化操作,模拟数据
19          visibleElements=new ArrayList<Element>();
20          allElements=new ArrayList<Element>();
21          //创建列表项信息,包括显示文字、所在层次、父节点 ID、是否包含子元素、是否展开
22          Element e1=new Element("食品饮料",Element.TOP_LEVEL,
                    0,Element.NO_PARENT,true,false);
23          Element e2=new Element("进口食品",Element.TOP_LEVEL+1,1,
                    e1.getId(),true,false);                 //添加第一层节点
24          Element e3=new Element("饼干蛋糕",Element.TOP_LEVEL+2,2,
                    e2.getId(),true,false);                 //添加第二层节点
25          Element e4=new Element("夹心饼干",Element.TOP_LEVEL+3,3,
                    e3.getId(),false,false);                //添加第三层节点
26          Element e5=new Element("地方特产",Element.TOP_LEVEL+1,4,
                    e1.getId(),true,false);                 //添加第一层节点
27          Element e6=new Element("西北区",Element.TOP_LEVEL+2,5,
                    e5.getId(),true,false);                 //添加第二层节点
28          Element e7=new Element("坚果类",Element.TOP_LEVEL+3,6,
                    e6.getId(),false,false);                //添加第三层节点
29          Element e8=new Element("休闲食品",Element.TOP_LEVEL+1,7,
                    e1.getId(),false,false);                //添加第一层节点
30          Element e9=new Element("家用电器",Element.TOP_LEVEL,
                    8,Element.NO_PARENT,true,false);        //添加最外层节点
31          Element e10=new Element("生活电器",Element.TOP_LEVEL+1,9,
```

```
                    e9.getId(),true,false);                //添加第一层节点
32      Element e11=new Element("取暖电器",Element.TOP_LEVEL+2,
                    10,e10.getId(),true,false);            //添加第二层节点
33      Element e12=new Element("暖手宝",Element.TOP_LEVEL+3,11,
                    e11.getId(),true,false);               //添加第三层节点
34      Element e13=new Element("美的牌",Element.TOP_LEVEL+4,12,
                    e12.getId(),false,false);              //添加第四层节点
35      //将列表项添加到集合中
36      allElements.add(e1);
37      allElements.add(e2);
38      allElements.add(e3);
39      allElements.add(e4);
40      allElements.add(e5);
41      allElements.add(e6);
42      allElements.add(e7);
43      allElements.add(e8);
44      allElements.add(e9);
45      allElements.add(e10);
46      allElements.add(e11);
47      allElements.add(e12);
48      allElements.add(e13);
49      //添加初始显示的元素
50      visibleElements.add(e1);
51      visibleElements.add(e9);
52    }
53    private class MyItemClickListener implements OnItemClickListener{
54        public void onItemClick(AdapterView<?>parent,View view,int position,
55            long id){
56            //获取单击的选项所代表的Element
57            Element element=(Element)myAdapter.getItem(position);
58            //判断单击的项有没有子项,如果没有,则不进行任何操作
                //如果有,则判断该项是处于展开状态还是处于关闭状态
59            if(!element.isHasChildren()){
60                return;
61            }
62            if(element.isExpanded()){
                //如果是由展开切换到关闭,则需要删除一些元素
63                element.setExpanded(false);
64                //删除节点内部对应的子节点数据,包括子节点的子节点
65                ArrayList<Element>elementsToDel=new ArrayList<Element>();
66                for(int i=position+1; i<visibleElements.size(); i++){
67                    //如果碰到和当前项同一级别的则退出循环
                    //否则把相关的元素都添加到需要删除的集合中
68                    if(element.getLevel()>=visibleElements.get(i).getLevel())
```

```
69                break;
70                elementsToDel.add(visibleElements.get(i));
71            }
72            visibleElements.removeAll(elementsToDel);
              //删除所有需要删除的元素
73            myAdapter.notifyDataSetChanged();
74        } else {            //如果是由关闭切换到展开,则需要添加一些元素
75            element.setExpanded(true);
76            //从数据源中提取子元素数据添加到列表
              //这里只是添加了下一级子节点
77            int i=1;        //注意这里的计数器放在for外面才能保证计数有效
78            //遍历所有的元素,如果元素的父ID与当前项的ID相同
              //则需要添加到显示的元素中
79            for(Element e : allElements){
80                if(e.getParendId()==element.getId()){
81                    e.setExpanded(false);
82                    visibleElements.add(position+i,e);
83                    i++;
84                }
85            }
86            myAdapter.notifyDataSetChanged();
87        }
88    }
89  }
90  private class MyAdapter extends BaseAdapter {  //自定义Adapter类
91      public int getCount(){                      //获取列表中包含的项数
92          return visibleElements.size();
93      }
94      public Object getItem(int position){        //获取指定项的对象
95          return visibleElements.get(position);
96      }
97      public long getItemId(int position){        //获取指定项的ID
98          return 0;
99      }
100     //获取指定项的控件
        public View getView(int position,View convertView,ViewGroup parent){
101         //将布局文件转换成View对象,用于显示每一项信息的控件
102         View view=getLayoutInflater().inflate(R.layout.item,null);
103         ImageView icon=(ImageView)view.findViewById(R.id.icon);
104         TextView text=(TextView)view.findViewById(R.id.text);
105         Element element=visibleElements.get(position);
            //获取当前位置的元素
106         int level=element.getLevel();
107         //设置图标的边距,主要是左边距,需要根据层次级别动态确定
```

```
108         icon.setPadding(basePadding * level,0,  0,0);
109         icon.setLayoutParams(new LinearLayout.LayoutParams(LayoutParams.
               WRAP_CONTENT,40-level*baseHeight));
110         text.setTextSize(TypedValue.COMPLEX_UNIT_PX,
               text.getTextSize()-level*baseSize);
111         text.setText(element.getText());
112         //显示图标状态,首先判断是否显示图片,然后判断显示什么图片
113         if(!element.isHasChildren()){    //如果该元素没有子元素,则不显示图片
114             icon.setImageResource(R.drawable.close);
115             icon.setVisibility(View.INVISIBLE);
116         } else {                         //表示该元素有子元素,显示图片
117             icon.setVisibility(View.VISIBLE);
118             if(element.isExpanded()){
                //判断该元素是否已经展开,如果展开则显示展开图片
119                 icon.setImageResource(R.drawable.open);
120             } else {                     //表示该元素没有展开,显示关闭图片
121                 icon.setImageResource(R.drawable.close);
122             }
123         }
124         return view;
125     }
126   }
127 }
```

15.3 代码分析

本案例通过 ListView 实现多级列表的效果,关键在于 Adapter 的构建和列表项的单击事件的处理。其本质上仍然是一级列表,只不过在 Adapter 的构建过程中根据该项所在的层次动态地设置它的左边距、图标大小、文字大小等,给人一种层次感,看起来就像是多级列表一样。然后在列表项的单击事件处理中判断该项是否包含子元素,如果包含子元素再继续判断该项是展开的还是折叠的,从而动态地从列表中删除项或者向列表中添加项。

列表中的项与项之间存在着一定的关系,这是以往的 ListView 所不具备的特点,项与项之间的关系通过 ID 来关联,除了顶层元素以外,列表中的每一项都有一个直接父元素,父元素与之有相同的结构。列表中每一项的具体信息包括文本内容、ID、父元素 ID、是否包含子元素、是否展开,在此通过 Element 类封装该信息,见 Element.java 的代码。

ListView 中显示的数据是通过 Adapter 指定的,由于此处需要根据列表项的状态来具体设置其显示方式,通过系统提供的 Adapter 无法实现该效果,只能通过自定义 Adapter 来实现。在自定义 Adapter 中,和列表项的显示有关的方法是 getView()。在该方法中,首先将列表项的布局文件转换成相应的 View,然后根据位置获取对应的 Element 对象,有了这个对象以后就可以知道该对象所在的层次,根据其层次结构来动态

地设置它的左边距、图标的大小、文字的大小等。然后判断其是否包含子元素，如果不包含子元素，则不显示图标；如果包含子元素，则显示图标，还需要继续判断具体显示哪个图标；如果该元素已经展开，则显示展开的图标，否则显示关闭的图标。代码见第100～125行。

在列表项的单击事件处理中，同样需要判断该元素是否包含子元素，如果不包含，则单击没有任何效果；如果包含子元素，则需要进一步判断单击时该项是处于展开还是关闭状态。如果是处于展开状态，则将其转化为关闭状态，同时不显示它的子元素；如果处于关闭状态，则将其转化为展开状态，同时显示它的子元素。通过单击能够在展开和关闭两种状态之间进行切换，同时列表中显示的数据也随之更新。

15.4 知识扩展

在上案例中列表项中的数据是在程序中临时模拟的，不能持久化保存，实际上，在大多数情况下，数据都保存在数据库中，特别是对于一些有增、删、改操作的应用。判断列表中的某项有没有子元素，也不是向上面那样直接给定，而应该通过查询数据库中有没有记录的parentId与当前元素的ID相同，如果有，则表明当前元素有子元素，否则表示该元素没有子元素。此外，判断列表中的某一项处于展开还是关闭状态，这是经常需要用到的，并且也是实时变化的数据，保存到数据库中意义不大，反而影响性能，在此让所有的项都默认是关闭的。

通过以上分析，下面采用数据库保存数据，然后从数据库中动态地获取相关数据显示在列表中，所有的布局文件不变，实现的功能相同。此时，Element.java的代码有所变化，具体如下。

自定义列表项：15\ProductCatagoriesExt\src\iet\jxufe\cn\android\productcatagoriesext\Element.java

```
1   public class Element {
2       private String text;           //文本信息内容
3       private int level;             //在层次结构中的级别，最顶级为0
4       private int id;                //该项元素的ID
5       private int parendId;          //直接父元素的ID，如果没有父元素该值为-1
6       private boolean isExpanded;    //该项是否展开
7       //定义两个常量，顶级元素level为0,父元素ID为-1
8       public static final int NO_PARENT=-1;
9       public static final int TOP_LEVEL=0;
10      public Element(String text,int level,int id,int parendId){  //构造方法
11          this.text=text;
12          this.level=level;
13          this.id=id;
14          this.parendId=parendId;
15      }
16      public Element(){}             //无参数的构造方法
```

```
17      //生成相应的set方法和get方法,设置和获取相应的属性值
18      public boolean isExpanded(){
19          return isExpanded;
20      }
21      public void setExpanded(boolean isExpanded){
22          this.isExpanded=isExpanded;
23      }
24      public String getText(){
25          return text;
26      }
27      public void setText(String text){
28          this.text=text;
29      }
30      public int getLevel(){
31          return level;
32      }
33      public void setLevel(int level){
34          this.level=level;
35      }
36      public int getId(){
37          return id;
38      }
39      public void setId(int id){
40          this.id=id;
41      }
42      public int getParendId(){
43          return parendId;
44      }
45      public void setParendId(int parendId){
46          this.parendId=parendId;
47      }
48  }
```

数据库辅助类:15\ProductCatagoriesExt\src\iet\jxufe\cn\android\productcatagoriesext\MyOpenHelper.java

```
1   public class MyOpenHelper extends SQLiteOpenHelper {
2       public String createTableSQL="create table if not exists element_tb"
3           +"(_id integer primary key autoincrement,id,text,level,parentId)";
4       public MyOpenHelper(Context context,String name,CursorFactory factory,
5           int version){
6           super(context,name,factory,version);
7       }
8       //数据库创建后回调该方法,执行建表操作和插入初始化数据的操作
9       public void onCreate(SQLiteDatabase db){
```

```
10          db.execSQL(createTableSQL);
11          init(db);
12      }
13      //数据库版本更新时回调该方法
14      public void onUpgrade(SQLiteDatabase db,int oldVersion,int newVersion){
15          System.out.println("版本变化:"+oldVersion+"-------->"+newVersion);
16      }
17      public void init(SQLiteDatabase db){
18          ArrayList<Element>list=new ArrayList<Element>();
19          Element e1=new Element("食品饮料",Element.TOP_LEVEL,
                    0,Element.NO_PARENT);
20          Element e2=new Element("进口食品",Element.TOP_LEVEL +1,1,
                    e1.getId());              //添加第一层结点
21          Element e3=new Element("饼干蛋糕",Element.TOP_LEVEL +2,2,
                    e2.getId());              //添加第二层结点
22          Element e4=new Element("夹心饼干",Element.TOP_LEVEL +3,3,
                    e3.getId());              //添加第三层结点
23          Element e5=new Element("地方特产",Element.TOP_LEVEL +1,4,
                    e1.getId());              //添加第一层结点
24          Element e6=new Element("西北区",Element.TOP_LEVEL +2,5,
                    e5.getId());              //添加第二层结点
25          Element e7=new Element("坚果类",Element.TOP_LEVEL +3,6,
                    e6.getId());              //添加第三层结点
26          Element e8=new Element("休闲食品",Element.TOP_LEVEL +1,7,
                    e1.getId());              //添加第一层结点
27          Element e9=new Element("家用电器",Element.TOP_LEVEL,
                    8,Element.NO_PARENT);     //添加最外层结点
28          Element e10=new Element("生活电器",Element.TOP_LEVEL +1,9,
                    e9.getId());              //添加第一层结点
29          Element e11=new Element("取暖电器",Element.TOP_LEVEL +2,
                    10,e10.getId());          //添加第二层结点
30          Element e12=new Element("暖手宝",Element.TOP_LEVEL +3,
                    11,e11.getId());          //添加第三层结点
31          Element e13=new Element("美的牌",Element.TOP_LEVEL +4,
                    12,e12.getId());          //添加第四层结点
32          list.add(e1);
33          list.add(e2);
34          list.add(e3);
35          list.add(e4);
36          list.add(e5);
37          list.add(e6);
38          list.add(e7);
39          list.add(e8);
40          list.add(e9);
```

```
41          list.add(e10);
42          list.add(e11);
43          list.add(e12);
44          list.add(e13);
45          for (Element element : list){
46              db.execSQL("insert into element_tb(id,text,level,parentId)
                    values(?,?,?,?)",
47                  new String[] { element.getId()+"",element.getText(),
                    element.getLevel()+"",element.getParendId()+""});
48          }
49      }
50  }
```

主程序代码：15\ProductCatagoriesExt\src\iet\jxufe\cn\android\productcatagoriesext\MainActivity.java

```
1   public class MainActivity extends Activity {
2       private ListView mListView;                        //列表控件
3       private MyAdapter myAdapter;                       //自定义 Adapter 对象
4       private ArrayList<Element>visibleElements;         //可见元素的集合
5       private int basePadding=20;                        //默认上层与下层左边距之间的差距
6       private int baseSize=2;                            //默认相邻级别字体相差 2px
7       private int baseHeight=8;                          //默认相邻级别图片高度相差 8dp
8       private MyOpenHelper mHelper;                      //数据库辅助类
9       private SQLiteDatabase mDB;                        //SQLite 数据库
10      protected void onCreate(Bundle savedInstanceState){
11          super.onCreate(savedInstanceState);
12          setContentView(R.layout.activity_main);
13          mListView= (ListView)findViewById(R.id.mListView);
14          mHelper=new MyOpenHelper(this,"element.db",null,1);
15          mDB=mHelper.getWritableDatabase();   //获取数据库
16          visibleElements=getData("select * from element_tb where parentId=?",
                new String[]{Element.NO_PARENT+""});
17          myAdapter=new MyAdapter();             //创建 Adapter
18          mListView.setAdapter(myAdapter);
19          mListView.setOnItemClickListener(new MyItemClickListener());
            //为列表项添加单击事件处理
20      }
21      private ArrayList<Element>getData(String sql,String[] args){
            //根据查询条件获取查询结果
22          ArrayList<Element>list=new ArrayList<Element>();
23          Cursor cursor=mDB.rawQuery(sql,args);
24          while(cursor.moveToNext()){
25              Element element=new Element();    //创建列表项对象
26              element.setId(cursor.getInt(cursor.getColumnIndex("id")));
```

```
27          element.setLevel(cursor.getInt(cursor.getColumnIndex("level")));
28          element.setParendId(cursor.getInt(cursor.getColumnIndex
                ("parentId"))); 
29          element.setText(cursor.getString(cursor.getColumnIndex
                ("text"))); 
30          element.setExpanded(false);        //默认每一项都是未展开的
31          list.add(element);                 //将列表项数据放入集合
32      }
33      return list;
34  }
35  private boolean hasChildren(Element element){ //判断某一项是否有子元素
36      ArrayList<Element>list=getData("select * from element_tb where
            parentId=?",new String[]{element.getId()+""});
37      if(list!=null&&list.size()!=0){
38          return true;
39      }else return false;
40  }
41  private class MyItemClickListener implements OnItemClickListener {
42      public void onItemClick(AdapterView<?>parent,View view,int position,
43              long id){
44          //获取单击的选项所代表的 Element
45          Element element=(Element)myAdapter.getItem(position);
46          //判断单击的项有没有子项,如果没有,则不进行任何操作
            //如果有,则判断该项是处于展开状态还是处于关闭状态
47          if(!hasChildren(element)){
48              return;
49          }
50          if(element.isExpanded()){
                //如果是由展开切换到关闭,则需要删除一些元素
51              element.setExpanded(false);
                //删除节点内部对应的子节点数据,包括子节点的子节点
53              ArrayList<Element>elementsToDel=new ArrayList<Element>();
54              for(int i=position +1; i <visibleElements.size(); i++){
55                  //如果碰到和当前项同一级别的则退出循环,否则把相关的
                    //元素都添加到需要删除的集合中
56                  if(element.getLevel()>=visibleElements.get(i).
                        getLevel())
57                      break;
58                  elementsToDel.add(visibleElements.get(i));
59              }
60              //删除所有需要删除的元素
61              visibleElements.removeAll(elementsToDel);
62              myAdapter.notifyDataSetChanged();
63          } else {             //如果是由关闭切换到展开,则需要添加一些元素
```

```
64                element.setExpanded(true);
65                //从数据库中查询该元素的子元素,并添加到列表中
                  //这里只查询直接子元素
66                ArrayList<Element>addElements=getData("select * from
                      element_tb where parentId=?",new String[]{element.
                      getId()+""});
67                visibleElements.addAll(position+1,addElements);
68                myAdapter.notifyDataSetChanged();
69            }
70        }
71    }
72    private class MyAdapter extends BaseAdapter { //自定义 Adapter 类
73        public int getCount(){                    //获取列表中包含的项数
74            return visibleElements.size();
75        }
76        public Object getItem(int position){   //获取指定项的对象
77            return visibleElements.get(position);
78        }
79        public long getItemId(int position){    //获取指定项的 ID
80            return 0;
81        }
82        //获取指定项的控件
83        public View getView(int position,View convertView,ViewGroup parent){
84            //将布局文件转换成 View 对象,用于显示每一项信息的控件
85            View view=getLayoutInflater().inflate(R.layout.item,null);
86            ImageView icon=(ImageView)view.findViewById(R.id.icon);
87            TextView text=(TextView)view.findViewById(R.id.text);
88            //获取当前位置的元素
89            Element element=visibleElements.get(position);
90            int level=element.getLevel();
91            //设置图标的边距,主要是左边距,需要根据层次级别动态确定
92            icon.setPadding(basePadding * level,0,  0,0);
93            icon.setLayoutParams(new LinearLayout.LayoutParams(LayoutParams.
                  WRAP_CONTENT,40-level*baseHeight));
94            text.setTextSize(TypedValue.COMPLEX_UNIT_PX,
                  text.getTextSize()-level*baseSize);
95            text.setText(element.getText());
96            //显示图标状态,首先判断是否显示图片,然后判断显示什么图片
97            if(!hasChildren(element)){   //如果该元素没有子元素,则不显示图片
98                icon.setImageResource(R.drawable.close);
99                icon.setVisibility(View.INVISIBLE);
100           } else {                    //表示该元素有子元素,显示图片
101               icon.setVisibility(View.VISIBLE);
102               if(element.isExpanded()){
```

```
                        //判断该元素是否已经展开,如果展开则显示展开图片
103                icon.setImageResource(R.drawable.open);
104            } else {                        //表示该元素没有展开,显示关闭图片
105                icon.setImageResource(R.drawable.close);
106            }
107        }
108        return view;
109    }
110 }
111 protected void onDestroy(){          //退出时关闭数据库连接
112     if(mDB!=null){
113         mDB.close();
114     }
115     super.onDestroy();
116 }
117 }
```

15.5 思考与练习

尝试为列表项添加长按事件处理,长按后弹出对话框,让用户选择操作,删除该项数据,为该项添加子项,在选择删除时,将该项及其子项删除,在选择添加子项时,可以设置子项的名称等。

第16章

天气预报——Web Service 的调用

16.1 案例概述

本案例主要实现天气预报功能,通过 Web Service 访问第三方提供的天气信息,获取到的是一连串的字符串信息,然后将这些信息解析、加工、设计成普通用户能一目了然的效果。其主要功能包括显示当前的天气实况、查看未来五天的天气信息、查看天气变化趋势图、切换城市、快速查看其他城市信息等,涉及 Web Service 调用、TabHost、Fragment、ListView、GridView、自定义 XML 图片等知识,程序运行效果如图 16-1 至图 16-6 所示。

图 16-1　程序运行效果图 1

图 16-2　程序运行效果图 2

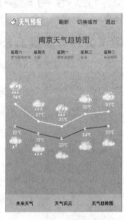

图 16-3　程序运行效果图 3

图 16-4　程序运行效果图 4

图 16-5　程序运行效果图 5

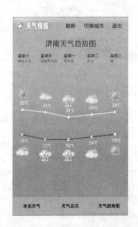

图 16-6　程序运行效果图 6

16.2 关键代码

主界面布局文件：16\WeatherForecast\res\layout\activity_main.xml

```xml
1   <LinearLayout xmlns:android="http://schemas.android.com/apk/res/android"
2       android:layout_width="match_parent"
3       android:layout_height="match_parent"
4       android:background="#aabbcc"
5       android:orientation="vertical" >
6       <!--显示"刷新"、"切换城市"、"退出"按钮 -->
7       <RelativeLayout
8           android:layout_width="match_parent"
9           android:layout_height="wrap_content"
10          android:background="@drawable/title_bg">
11          <Button
12              android:id="@+id/exit"
13              android:text="@string/exit"
14              style="@style/btnStyle"
15              android:layout_alignParentRight="true"
16              android:onClick="exit"/>
17          <Button
18              android:id="@+id/changeCity"
19              android:layout_toLeftOf="@id/exit"
20              android:text="@string/changeCity"
21              style="@style/btnStyle"
22              android:onClick="changeCity"/>
23          <Button
24              android:layout_toLeftOf="@id/changeCity"
25              android:text="@string/refresh"
26              style="@style/btnStyle"
27              android:onClick="refresh"/>
28      </RelativeLayout>
29      <!--显示选项卡信息,选项和切换的页面 -->
30      <TabHost
31          android:id="@+id/mTabHost"
32          android:layout_width="match_parent"
33          android:layout_height="match_parent" >
34          <LinearLayout
35              android:layout_width="match_parent"
36              android:layout_height="match_parent"
37              android:orientation="vertical" >
38              <FrameLayout
39                  android:id="@android:id/tabcontent
```

```
40                  android:layout_width="match_parent"
41                  android:layout_height="0dp"
42                  android:layout_weight="1" >
43              <FrameLayout
44                  android:id="@+id/realContent"
45                  android:layout_width="match_parent"
46                  android:layout_height="match_parent" />
47              </FrameLayout>
48              <TabWidget
49                  android:id="@android:id/tabs"
50                  android:layout_width="match_parent"
51                  android:layout_height="wrap_content"
52                  android:background="#8800cccc" />
53          </LinearLayout>
54      </TabHost>
55  </LinearLayout>
```

程序主界面：16\WeatherForecast\src\iet\jxufe\cn\android\weatherforecast\MainActivity.java

```
1   public class MainActivity extends Activity {
2       public static SoapObject weatherObject;        //天气信息对象
3       public static String province,city;            //获取保存的省份、城市信息
4       private SharedPreferences mPreferences;        //获取参数信息
5       private SharedPreferences.Editor mEditor;      //参数编辑器
6       private TabHost weatherTab;                    //选项卡
7       private int count=1;                           //记录单击退出的次数
8       private int currentTab=1;                      //当前的选项页,默认为1,即第二项
9       protected void onCreate(Bundle savedInstanceState){
10          super.onCreate(savedInstanceState);
11          requestWindowFeature(Window.FEATURE_NO_TITLE);   //去除标题栏
12          getWindow().setFlags(WindowManager.LayoutParams.FLAG_FULLSCREEN,
13              WindowManager.LayoutParams.FLAG_FULLSCREEN);//全屏显示
14          setContentView(R.layout.activity_main);
15          mPreferences=getSharedPreferences("weather",Context.MODE_PRIVATE);
                //获取参数信息
16          province=mPreferences.getString("province","江西");
                //获取保存的省份信息
17          city=mPreferences.getString("city","南昌");
                //获取保存的城市信息,默认为江西—南昌
18          weatherTab= (TabHost)findViewById(R.id.mTabHost);   //获取选项卡
19          weatherTab.setup();
20          //选项卡的初始化,添加3个简单选项
21          weatherTab.addTab(weatherTab.newTabSpec("weatherTrend")
22              .setIndicator("未来天气").setContent(R.id.realContent));
```

```
23          weatherTab.addTab(weatherTab.newTabSpec("currentWeather")
24                  .setIndicator("天气实况").setContent(R.id.realContent));
25          weatherTab.addTab(weatherTab.newTabSpec("trendChart")
26                  .setIndicator("天气趋势图").setContent(R.id.realContent));
27          getWeatherInfo(city);                      //调用方法,获取天气信息
28      }
29      private Handler mHandler=new Handler(){
             //创建 Handler 对象发送、接收、处理消息
30          public void handleMessage(Message msg){
31              if(msg.what==0x11){
32                  if(weatherObject !=null){
                         //判断天气信息是否为空,若不为空显示信息
33                      weatherTab.setOnTabChangedListener(new MyTabListener());
34                      weatherTab.setCurrentTab((currentTab+1)%3);       //切换页面
35                      weatherTab.setCurrentTab((currentTab+3 -1)%3); //恢复页面
36                  } else {       //如果未能获取到天气信息,则提示用户查看网络状态
37                      Toast.makeText(MainActivity.this,"无法获取天气信息,
                         请检查网络状态!",
                         Toast.LENGTH_SHORT).show();
38                  }
39              }
40          }
41      };
42      public void getWeatherInfo(final String city){
             //启动一个线程,向 Web Service 发送请求,获取天气信息
43          new Thread(){                              //创建线程
44              public void run(){
45                  weatherObject=WebServiceUtil.getWeatherByCity(city);
                     //调用辅助类的方法,根据城市获取信息
46                  mHandler.sendEmptyMessage(0x11);   //获取到信息后发送消息
47              }
48          }.start();                                 //启动线程
49      }
50      private class MyTabListener implements OnTabChangeListener {    //页面监听器
51          public void onTabChanged(String tabTag){
                 //TabHost 中的页面切换时调用该方法
52              currentTab=weatherTab.getCurrentTab();     //保存当前页面序号
53              FragmentTransaction fTransaction=getFragmentManager()
54                  .beginTransaction();   //启动事务,判断当前用户切换到哪个页面
55              if(tabTag.equalsIgnoreCase("currentWeather")){
                     //天气实况页面,加载相关信息
56                  fTransaction.replace(R.id.realContent,new
                         WeatherInfoFragment());
57              } else if(tabTag.equals("weatherTrend")){
```

```
                        //未来天气,未来五天天气信息
58            fTransaction.replace(R.id.realContent,new
                  WeatherListFragment());
59        } else if(tabTag.equalsIgnoreCase("trendChart")){
              //天气趋势图,绘制天气趋势
60            fTransaction.replace(R.id.realContent,new
                  WeatherTrendFragment());
61        }
62        fTransaction.commit();              //提交事务
63    }
64  }
65  protected void onActivityResult(int requestCode,int resultCode,
        Intent data){
        //获取用户选择的城市
66      if(requestCode==0 && resultCode==0){
67          if(data !=null){                        //如果返回的数据不为空
68              city=data.getStringExtra("city");//获取城市
69              province=data.getStringExtra("province");  //获取省份信息
70              mEditor=mPreferences.edit();        //得到参数编辑器
71              mEditor.putString("city",city);     //保存变化的城市信息
72              mEditor.putString("province",province);   //保存变化的省份信息
73              mEditor.commit();                   //提交变化
74              refresh(null);                      //刷新页面
75          }
76      }
77  }
78  public void exit(View view){         //退出按钮事件处理,连续按两次退出应用
79      if(count < 2){                   //如果是第一次,则提示用户再按一次
80          Toast.makeText(this,"再按一次退出天气预报",
                Toast.LENGTH_SHORT).show();
81          count++;
82          new Thread(){                //创建线程用于计时,3s 后恢复原状
83              public void run(){
84                  try {
85                      sleep(3000);     //3s 以后恢复为原值
86                      count=1;
87                  } catch(Exception e){
88                      e.printStackTrace();
89                  }
90              }
91          }.start();
92      } else {                         //如果连按两次,则退出
93          this.finish();
94      }
```

```
95        }
96     public void refresh(View view){    //刷新按钮事件处理,重新获取天气信息
97         getWeatherInfo(city);
98     }
99     public void changeCity(View view){
           //切换城市按钮事件处理,跳转到选择城市页面
100        Intent intent=new Intent(this,ChooseCityActivity.class);
101        startActivityForResult(intent,0);
102    }
103    public boolean onKeyDown(int keyCode,KeyEvent event){
           //按返回键的事件处理
104        if(keyCode==KeyEvent.KEYCODE_BACK){
105            exit(null);
106            return true;
107        }
108        return super.onKeyDown(keyCode,event);
109    }
110 }
```

调用 Web Service 的辅助类:16\WeatherForecast\src\iet\jxufe\cn\android\weatherforecast\WebServiceUtil.java

```
1  public class WebServiceUtil {                //调用天气预报的 Web Service 工具类
2     public static final String SERVICE_URL="http://webservice.webxml.com.
           cn/WebServices/WeatherWS.asmx";  //服务地址
3     public static final String SERVICE_NAMESPACE="http://WebXml.com.cn/";
       //服务的命名空间需要和 WSDL 文档中的一致
4     public static List<String>getProvinces(){  //获取提供天气信息的所有省份
5        List<String>provinces=new ArrayList<String>();    //保存获取到的省份信息
6        String method="getRegionProvince";                //需要调用的方法名
7        //创建 Soap 对象(简单对象访问协议),传递两个参数
         //第一个为服务的命名空间,第二个为需要调用的方法名
8        SoapObject soapObject=new SoapObject(SERVICE_NAMESPACE,method);
9        //创建 SoapSerializationEnvelope 对象,传递 Soap 的版本号,在此使用 SOAP 12
10       SoapSerializationEnvelope mEnvelope=new SoapSerializationEnvelope
                                    (SoapEnvelope.VER12);
11       mEnvelope.bodyOut=soapObject;
12       mEnvelope.dotNet=true;
13       HttpTransportSE httpsTransportSE=new HttpTransportSE(SERVICE_URL);
14       try{
15           httpsTransportSE.call(SERVICE_NAMESPACE+method, mEnvelope);
16           if(mEnvelope.getResponse()!=null){
17               SoapObject result=(SoapObject)mEnvelope.bodyIn;
18               SoapObject detail=(SoapObject)result.getProperty
                     (method+"Result");
```

```java
19              for(int i=0;i<detail.getPropertyCount();i++){
20                  provinces.add(detail.getProperty(i).toString().
                        split(",")[0]);
21              }
22          }
23      }catch(Exception ex){
24          ex.printStackTrace();
25      }
26      return provinces;
27  }
28  public static List<String>getCitiesByProvince(String province){
        //根据省份查询城市
29      List<String>cities=new ArrayList<String>();
30      HttpTransportSE httpTransportSE=new HttpTransportSE(SERVICE_URL);
31      SoapSerializationEnvelope mEnvelope=new SoapSerializationEnvelope
                                    (SoapEnvelope.VER12);
32      mEnvelope.dotNet=true;
33      String method="getSupportCityString";        //方法名
34      SoapObject soapObject=new SoapObject(SERVICE_NAMESPACE,method);
35      soapObject.addProperty("theRegionCode",province);
36      mEnvelope.bodyOut=soapObject;
37      try{
38          httpTransportSE.call(SERVICE_NAMESPACE+method, mEnvelope);
39          if(mEnvelope.getResponse()!=null){
40              SoapObject result=(SoapObject)mEnvelope.bodyIn;
41              SoapObject detail=(SoapObject)result.getProperty
                    (method+"Result");
42              for(int i=0;i<detail.getPropertyCount();i++){
43                  cities.add(detail.getProperty(i).toString().
                        split(",")[0]);
                    //获取城市并添加到集合中
44              }
45          }
46      }catch(Exception ex){
47          ex.printStackTrace();
48      }
49      return cities;
50  }
51  public static SoapObject getWeatherByCity(String city){//根据城市查询天气
52      HttpTransportSE httpTransportSE=new HttpTransportSE(SERVICE_URL);
53      httpTransportSE.debug=true;
54      SoapSerializationEnvelope mEnvelope=new SoapSerializationEnvelope
                                    (SoapEnvelope.VER12);
55      mEnvelope.dotNet=true;
```

```
56        String method="getWeather";
57        SoapObject soapObject=new SoapObject(SERVICE_NAMESPACE,method);
58        soapObject.addProperty("theCityCode",city);
59        mEnvelope.bodyOut=soapObject;
60        try{
61            httpTransportSE.call(SERVICE_NAMESPACE+method, mEnvelope);
62            if(mEnvelope.getResponse()!=null){
63                SoapObject result=(SoapObject)mEnvelope.bodyIn;
64                SoapObject detail=(SoapObject)result.getProperty
                        (method+"Result");
65                return detail;
66            }
67        }catch(Exception ex){
68            ex.printStackTrace();
69        }
70        return null;
71    }
72 }
```

显示天气实况信息的布局文件：16\WeatherForecast\res\layout\weather_info.xml

```
1  <LinearLayout xmlns:android="http://schemas.android.com/apk/res/android"
2      android:layout_width="match_parent"
3      android:layout_height="match_parent"
4      android:background="@drawable/body_bg"
5      android:orientation="vertical">
6      <RelativeLayout
7          android:layout_width="match_parent"
8          android:layout_height="wrap_content"
9          android:layout_margin="5dp"
10         android:background="@drawable/weather_boder"
11         android:padding="5dp">
12         <TextView
13             android:text="@string/currentWeather"
14             android:layout_centerHorizontal="true"
15             android:textSize="16sp"
16             android:textStyle="bold"
17             android:textColor="#0000ff"
18             style="@style/textStyle"/>
19         <TextView
20             android:id="@+id/city"
21             android:layout_width="wrap_content"
22             android:layout_height="wrap_content"
23             android:layout_marginLeft="10dp"
```

```
24          android:paddingTop="10dp"
25          android:textSize="16sp"/>
26      <DigitalClock
27          android:id="@+id/time"
28          android:layout_width="wrap_content"
29          android:layout_height="wrap_content"
30          android:layout_alignLeft="@id/city"
31          android:layout_below="@id/city"
32          android:textSize="20sp"/>
33      <TextView
34          android:id="@+id/date"
35          android:layout_width="wrap_content"
36          android:layout_height="wrap_content"
37          android:layout_alignLeft="@id/city"
38          android:layout_below="@id/time"
39          android:padding="5dp"
40          android:textSize="12sp"/>
41      <TextView
42          android:id="@+id/refreshTime"
43          android:layout_width="wrap_content"
44          android:layout_height="wrap_content"
45          android:layout_alignLeft="@id/city"
46          android:layout_below="@id/date"
47          android:drawableLeft="@drawable/refresh"
48          android:gravity="center_vertical"
49          android:paddingBottom="10dp"
50          android:textSize="12sp"
51          android:textStyle="bold"/>
52      <LinearLayout android:layout_width="wrap_content"
53          android:layout_height="wrap_content"
54          android:orientation="vertical"
55          android:layout_marginRight="30dp"
56          android:paddingTop="10dp"
57          android:layout_alignParentRight="true"
58          android:layout_centerVertical="true">
59          <TextView
60              android:id="@+id/temperView"
61              android:textSize="16sp"
62              style="@style/textStyle"/>
63          <TextView
64              android:id="@+id/windDirView"
65              style="@style/textStyle"/>
66          <TextView
67              android:id="@+id/windPowerView"
```

```
68              style="@style/textStyle"/>
69          <TextView
70              android:id="@+id/humidityView"
71              style="@style/textStyle"/>
72          </LinearLayout>
73      </RelativeLayout>
74      <TextView
75          android:layout_width="match_parent"
76          android:layout_height="wrap_content"
77          android:gravity="center"
78          android:text="@string/hint"
79          android:textSize="20sp"/>
80      <ListView
81          android:id="@+id/weatherHints"
82          android:layout_width="match_parent"
83          android:layout_height="wrap_content"/>
84  </LinearLayout>
```

布局中涉及的样式信息：16\WeatherForecast\res\values\styles.xml

```
1   <resources xmlns:android="http://schemas.android.com/apk/res/android">
2       <style name="btnStyle">
3           <item name="android:layout_width">wrap_content</item>
4           <item name="android:layout_height">wrap_content</item>
5           <item name="android:textSize">14sp</item>
6           <item name="android:background">@drawable/btn_bg</item>
7       </style>
8       <style name="textStyle">
9           <item name="android:layout_width">wrap_content</item>
10          <item name="android:layout_height">wrap_content</item>
11          <item name="android:textSize">14sp</item>
12          <item name="android:paddingBottom">5dp</item>
13      </style>
14  </resources>
```

显示天气实况信息的页面 Fragment：
16\WeatherForecast\src\iet\jxufe\cn\android\weatherforecast\WeatherInfoFragment.java

```
1   public class WeatherInfoFragment extends Fragment {    //用于显示当前天气信息的页面
2       private ListView weatherHints;                      //当前天气下的温馨提示列表
3       private TextView cityView,dateView,refreshTimeView,temperView,
4           windDirView,windPowerView,humidityView;
            //显示城市、日期、刷新时间、温度、风向、风力、湿度
5       private SoapObject weatherObject;                    //天气对象
```

```
6       private String[] hints;                              //提示信息
7       private List<Map<String,Object>>list=new ArrayList<Map<String,
            Object>>();
        //保存提示信息集合
8       public void onCreate(Bundle savedInstanceState){
9           super.onCreate(savedInstanceState);
10          weatherObject=MainActivity.weatherObject;
11          hints=weatherObject.getProperty(6).toString().split("\n");
        //通过换行符进行分割
12          for (int i=0; i<hints.length; i++){
13              Map<String,Object>item=new HashMap<String,Object>();
14              item.put("hintName",hints[i].split(":")[0]+":");
                //通过冒号分割提示信息,注意此处是中文的冒号
15              item.put("hintInfo",hints[i].split(":")[1]);
16              list.add(item);
17          }
18      }
19      public View onCreateView(LayoutInflater inflater,ViewGroup container,
20              Bundle savedInstanceState){
21          View infoView=inflater.inflate(R.layout.weather_info,
                container,false);
22          cityView=(TextView)infoView.findViewById(R.id.city);
23          dateView=(TextView)infoView.findViewById(R.id.date);
24          temperView=(TextView)infoView.findViewById(R.id.temperView);
25          windDirView=(TextView)infoView.findViewById(R.id.windDirView);
26          windPowerView=(TextView)infoView.findViewById(R.id.windPowerView);
27          humidityView=(TextView)infoView.findViewById(R.id.humidityView);
28          refreshTimeView=(TextView)infoView.findViewById(R.id.refreshTime);
29          cityView.setText(MainActivity.city);           //显示当前城市
30          Calendar calendar=Calendar.getInstance();      //获取当前时间
31          dateView.setText(calendar.get(Calendar.YEAR)+"-"
32              +(calendar.get(Calendar.MONTH)+1)+"-"
33              +calendar.get(Calendar.DAY_OF_MONTH)+" "
34              +Util.intToWeek(calendar.get(Calendar.DAY_OF_WEEK)));
                //显示年月日以及星期信息
35          refreshTimeView.setText("今天:"
36              +weatherObject.getProperty(3).toString().split(" ")[1]);
                //显示刷新时间
37          String info=weatherObject.getProperty(4).toString();  //获取天气实况信息
38          String[] infos=info.split(";");                //每项信息通过分号隔开
39          temperView.setText(Html.fromHtml("温度:<font color=blue><b><i>"+
                infos[0].split(":")[2]+"</i></b></font>"));   //温度信息
40          String windStr=infos[1].split(":")[1];         //获取风向/风力信息
41          windDirView.setText(Html.fromHtml("风向:<font color=blue><b>"+
```

```
41            windStr.split("")[0]+"</b></font>"));    //风向信息
42            windPowerView.setText(Html.fromHtml("风力:<font color=blue><b>"+
                  windStr.split("")[1]+"</b></font>"));    //风力信息
43            humidityView.setText(Html.fromHtml("湿度:<font color=blue><i>"+
                  infos[2].split(":")[1]+"</i></font>"));    //湿度信息
44            weatherHints=(ListView)infoView.findViewById(R.id.weatherHints);
45            SimpleAdapter adapter=new SimpleAdapter(getActivity(),list,
46                R.layout.text_item,new String[] { "hintName","hintInfo" },
47                new int[] { R.id.hintName,R.id.hintInfo });
48            weatherHints.setAdapter(adapter);
49            return infoView;
50        }
51  }
```

显示未来天气信息的布局文件：16\WeatherForecast\res\layout\weather_list.xml

```
1   <LinearLayout xmlns:android="http://schemas.android.com/apk/res/android"
2       android:layout_width="wrap_content"
3       android:layout_height="wrap_content"
4       android:orientation="vertical"
5       android:background="@drawable/body_bg">
6       <TextView
7           android:id="@+id/futureTitle"
8           android:layout_width="match_parent"
9           android:layout_height="wrap_content"
10          android:gravity="center"
11          android:textSize="20sp"
12          android:padding="10dp"/>
13      <ListView
14          android:id="@+id/weatherList"
15          android:layout_height="wrap_content"
16          android:layout_width="wrap_content"
17          android:divider="#00000000"
18          android:dividerHeight="20dp"
19          android:padding="10dp"/>
20  </LinearLayout>
```

每一项天气信息的布局文件：16\WeatherForecast\res\layout\item.xml

```
1   <RelativeLayout xmlns:android="http://schemas.android.com/apk/res/android"
2       android:layout_width="match_parent"
3       android:layout_height="match_parent">
4       <RelativeLayout
5           android:layout_width="wrap_content"
```

```
6        android:layout_height="wrap_content"
7        android:layout_centerInParent="true"
8        android:padding="10dp"
9        android:layout_marginLeft="40dp"
10       android:background="@drawable/weather_boder" >
11     <ImageView
12         android:id="@+id/icon1"
13         android:layout_width="wrap_content"
14         android:layout_height="wrap_content"
15         android:layout_centerVertical="true"
16         android:layout_margin="5dp"
17         android:contentDescription="@string/imageInfo"/>
18     <ImageView
19         android:id="@+id/icon2"
20         android:layout_width="wrap_content"
21         android:layout_height="wrap_content"
22         android:layout_centerVertical="true"
23         android:layout_margin="5dp"
24         android:layout_toRightOf="@id/icon1"
25         android:contentDescription="@string/imageInfo"/>
26     <LinearLayout
27         android:layout_width="wrap_content"
28         android:layout_height="wrap_content"
29         android:layout_centerVertical="true"
30         android:layout_toRightOf="@id/icon2"
31         android:orientation="vertical" >
32         <TextView
33             android:id="@+id/date"
34             style="@style/textStyle"/>
35         <TextView
36             android:id="@+id/weather"
37             style="@style/textStyle"/>
38         <TextView
39             android:id="@+id/temper"
40             style="@style/textStyle"/>
41         <TextView
42             android:id="@+id/wind"
43             style="@style/textStyle"/>
44     </LinearLayout>
45   </RelativeLayout>
46 </RelativeLayout>
```

显示未来五天天气信息的页面 Fragment：
16\WeatherForecast\src\iet\jxufe\cn\android\weatherforecast\WeatherInfoFragment.java

```java
1   public class WeatherListFragment extends Fragment {
2       private ListView weatherList;                          //未来五天的天气情况
3       private SoapObject weatherObject;
4       private TextView futureTitle;
5       private List<Map<String,Object>>list=new ArrayList<Map<String,Object>>();
6       private SimpleAdapter adapter;
7       public void onCreate(Bundle savedInstanceState){
8           super.onCreate(savedInstanceState);
9           weatherObject=MainActivity.weatherObject;
10          init();                                             //执行初始化操作
11          adapter=new SimpleAdapter(getActivity(),list,R.layout.item,
12              new String[]{"date","weather","temper","wind","icon1","icon2"},
                new int[]{
13              R.id.date,R.id.weather,R.id.temper,R.id.wind,R.id.icon1,
                    R.id.icon2});
14      }
15      public View onCreateView(LayoutInflater inflater,ViewGroup container,
16              Bundle savedInstanceState){
17          View weahterListView=inflater.inflate(R.layout.weather_list,
18              container,false);
19          futureTitle=(TextView)weahterListView.findViewById
                (R.id.futureTitle);                             //标题
20          weatherList=(ListView)weahterListView.findViewById
                (R.id.weatherList);
21          futureTitle.setText(Html.fromHtml("<font color=blue><b>"+
                MainActivity.city+"</b></font>未来五天天气状况"));
22          weatherList.setAdapter(adapter);
23          return weahterListView;
24      }
25      public void init(){
26          for (int i=0;i<5;i++){                              //获取未来五天的天气信息
27              Map<String,Object>item=new HashMap<String,Object>();
28              String dateInfo=weatherObject.getProperty(7+5 * i).
                    toString();                                 //日期信息和天气
29              String date=dateInfo.split(" ")[0];             //获取日期信息
30              String weather=dateInfo.split(" ")[1];          //获取天气信息
31              String temperInfo=weatherObject.getProperty(7+5 * i+1).
                    toString();                                 //温度信息
32              String windInfo=weatherObject.getProperty(7+5 * i+2).
                    toString();                                 //风向信息
```

```
33          String icon1Str=weatherObject.getProperty(7+5*i+3).
                toString();                                    //天气图标1
34          String icon2Str=weatherObject.getProperty(7+5*i+4).
                toString();                                    //天气图标2
35          item.put("date","日期:"+date);
36          item.put("weather","天气:"+weather);
37          item.put("temper","温度:"+temperInfo);
38          item.put("wind","风向:"+windInfo);
39          item.put("icon1",Util.nameToImageId(icon1Str,"b_"));
40          item.put("icon2",Util.nameToImageId(icon2Str,"b_"));
41          list.add(item);
42      }
43   }
44 }
```

显示天气趋势的页面布局文件：16\WeatherForecast\res\layout\trend.xml

```
1  <LinearLayout xmlns:android="http://schemas.android.com/apk/res/android"
2      android:layout_width="match_parent"
3      android:layout_height="match_parent"
4      android:background="@drawable/body_bg"
5      android:orientation="vertical">
6      <TextView
7          android:id="@+id/trendTitle"
8          android:layout_width="match_parent"
9          android:layout_height="wrap_content"
10         android:gravity="center"
11         android:textSize="20sp"
12         android:padding="10dp"/>
13     <iet.jxufe.cn.android.weatherforecast.WeatherTrendView
14         android:layout_height="match_parent"
15         android:layout_width="match_parent"
16         android:background="@drawable/body_bg"/>
17 </LinearLayout>
```

显示天气趋势图 Fragment：
16\WeatherForecast\src\iet\jxufe\cn\android\weatherforecast\WeatherTrendFragment.java

```
1  public class WeatherTrendFragment extends Fragment {
2      public View onCreateView(LayoutInflater inflater, ViewGroup container,
3              Bundle savedInstanceState){
4          View trendView=inflater.inflate(R.layout.trend, null, false);
5          TextView title=(TextView)trendView.findViewById(R.id.trendTitle);
6          title.setText(Html.fromHtml("<font color=blue><b>"+
```

```
                MainActivity.city+"</b>
            </font>天气趋势图"));
7       return trendView;
8   }
9 }
```

自定义趋势图 View：16\WeatherForecast\src\iet\jxufe\cn\android\weatherforecast\WeatherTrendView.java

```
1  public class WeatherTrendView extends View{   //绘制天气变化趋势图
2      private Context context;
3      private Paint pointPaint,textPaint,textPaint2,linePaint;
       //坐标点画笔、文字画笔、线画笔
4      private float textHeight;              //文字高度
5      private int scale=10;                  //一度对应多少像素
6      private int radius=5;                  //温度坐标点的半径
7      private int xSpace=60;                 //横坐标点之间的间隔
8      private int ySpace=5;                  //温度文字与图标之间的垂直间隔
9      private int[] x=new int[5];            //一共有多少个温度值
10     private String[] weekDay=new String[5];  //显示的星期信息
11     private SoapObject weatherObject;      //天气对象
12     private List<Map<String,Object>>weatherInfoList=new ArrayList
           <Map<String,Object>>();
       //保存天气信息的列表
13     public WeatherTrendView(Context context){
14         this(context,null);
15     }
16     public WeatherTrendView(Context context,AttributeSet attrs){
17         super(context,attrs);
18         this.context=context;
19         init();
20     }
21     public void init(){                    //执行初始化操作
22         //坐标点画笔的初始化
23         pointPaint=new Paint();
24         pointPaint.setAntiAlias(true);
25         pointPaint.setColor(Color.WHITE);  //白色
26         //连接线画笔的初始化
27         linePaint=new Paint();
28         linePaint.setColor(Color.YELLOW);  //连接线的颜色为黄色
29         linePaint.setAntiAlias(true);
30         linePaint.setStrokeWidth(3);       //宽度为 3px
31         linePaint.setStyle(Style.FILL);    //填充
32         //文本画笔的初始化
33         textPaint=new Paint();
```

```
34        textPaint.setAntiAlias(true);
35        textPaint.setColor(Color.BLACK);        //文本颜色为黑色
36        textPaint.setTextSize(16);              //文字大小为16
37        textPaint2=new Paint();
38        textPaint2.setAntiAlias(true);
39        textPaint2.setColor(Color.RED);         //文本颜色为红色
40        textPaint2.setTextSize(14);             //文字大小为14
41        //计算文字高度
42        FontMetrics fontMetrics=textPaint.getFontMetrics();
43        textHeight=fontMetrics.bottom-fontMetrics.top;
          //文字底部坐标减去顶部坐标
44        getWeatherInfo();                                       //获取天气信息
45        Calendar calendar=Calendar.getInstance();               //获取当前时间
46        int dayOfWeek=calendar.get(Calendar.DAY_OF_WEEK);       //获取当前星期
47        for(int i=0;i<weekDay.length;i++){
48            weekDay[i]=Util.intToWeek((dayOfWeek+i)%7+1);
              //初始化未来五天对应的星期
49        }
50    }
51    protected void onDraw(Canvas canvas){       //在界面中进行绘制
52        super.onDraw(canvas);
53        int width=this.getWidth();              //获取控件的宽度
54        xSpace=width/x.length;                  //计算两个点之间的X轴间距
55        for(int i=0;i<x.length;i++){            //初始化各个点的X轴坐标,起始点偏移20px
56            x[i]=40+i*xSpace;
57        }
58        for(int i=0;i<weekDay.length;i++){
59            canvas.drawText(weekDay[i],x[i]-20,20,textPaint);
60            canvas.drawText((String)weatherInfoList.get(i).get
                  ("weather"),x[i]-20,20+textHeight,textPaint2);
61        }
62        textPaint.setTextSize(18);
63        textPaint.setColor(Color.WHITE);
64        int mindTem= (getMaxTem(weatherInfoList)+
              getMinTem(weatherInfoList))/2;//获取最高温度与最低温度的中间值
65        int centerHeight=this.getHeight()/2;
          //获取控件的Y轴中线,中线为中间温度,比该温度高的在上方,比该温度低的在下方
66        //绘制最高温度曲线
67        for(int i=0;i<weatherInfoList.size();i++){      //依次获取每一个温度值
68            int topTem= (Integer) weatherInfoList.get(i).get("topTem");
69            //该温度相对于中间温度的偏移量,然后根据温度差来计算它的位置,一度对应
              //scale个像素,如果为负值表示在中线上方,如果为正值表示在中线下方
70            float point= (-(topTem-mindTem))*scale;     //该点相对于中线的纵坐标
```

```
71        canvas.drawCircle(x[i],centerHeight+point,radius,
              pointPaint);                              //绘制坐标点
72        canvas.drawText(topTem+"℃",x[i]-12,centerHeight+point-
              textHeight/2-ySpace,textPaint);   //绘制该点对应的文字信息
73        //绘制该点对应的图片
74        int topPicId=(Integer) weatherInfoList.get(i).get("icon1");
          //获取图片的 ID
75        Bitmap bitmap=BitmapFactory.decodeResource(
76            context.getResources(),topPicId);  //根据 ID 解析得到位图 Bitmap
77        canvas.drawBitmap(bitmap,x[i]-bitmap.getWidth()/2,
              centerHeight+point-textHeight-2*ySpace-
              bitmap.getHeight(),null);          //在画布上绘制图片
78        if(i !=(weatherInfoList.size()-1)){
          //如果该点不是最后一个点,则需要绘制该点到下一个点的连线
79            int nextTopTem= (Integer) weatherInfoList.get(i+1).
                  get("topTem");
              //获取下一个温度值
80            float pointNext=(-(nextTopTem-mindTem)) * scale;
              //该温度相对于中线的偏移量
81            canvas.drawLine(x[i],centerHeight+point,x[i+1],
                  centerHeight+pointNext,linePaint);
82        }
83    }
84    linePaint.setColor(Color.BLUE);
85    //绘制最低温度曲线
86    for(int i=0;i<weatherInfoList.size();i++){        //依次获取每一个温度值
87        int lowTem=(Integer) weatherInfoList.get(i).get("lowTem");
88        float point= (-(lowTem-mindTem)) * scale;
          //该温度相对于中线的偏移量
89        canvas.drawCircle(x[i],centerHeight+point,radius,
              pointPaint);                              //绘制坐标点
90        canvas.drawText(lowTem+"℃",x[i]-12,centerHeight+point+
              textHeight+ySpace,textPaint);    //绘制坐标点对应的文字信息
91        int topPicId=(Integer) weatherInfoList.get(i).get("icon2");
          //获取图片 ID
92        Bitmap bitmap=BitmapFactory.decodeResource(
93            context.getResources(),topPicId);   //根据 ID 构建 Bitmap 对象
94            canvas. drawBitmap (bitmap, x [i] - bitmap. getWidth ( )/2,
              centerHeight+point+textHeight+2*ySpace,null);   //绘制图片
95        if(i !=(weatherInfoList.size()-1)){
              //如果该点不是最后一个点,则需要绘制该点到下一个点的连线
96            int nextLowTem= (Integer) weatherInfoList.get(i+1).
                  get("lowTem");
```

```
97                    float pointNext=(-(nextLowTem-mindTem))*scale;
98                    canvas.drawLine(x[i],centerHeight+point,x[i+1],
99                        centerHeight+pointNext,linePaint);
100               }
101          }
102      }
103      public int getMaxTem(List<Map<String,Object>>weatherInfoList){
         //获取最高温度
104          int temp=(Integer)weatherInfoList.get(0).get("topTem");
             //默认第一个为最高温度
105          for(int i=0;i<weatherInfoList.size();i++){
                 //循环遍历,如果有更高的则更改
106              if(temp<(Integer)weatherInfoList.get(i).get("topTem")){
107                  temp=(Integer)weatherInfoList.get(i).get("topTem");
108              }
109          }
110          return temp;
111      }
112      public int getMinTem(List<Map<String,Object>>weatherInfoList){
         //获取最低温度
113          int temp=(Integer)weatherInfoList.get(0).get("lowTem");
             //默认第一个为最低温度
114          for(int i=0;i<weatherInfoList.size();i++){
                 //循环遍历,如果有更低的则更改
115              if(temp>(Integer)weatherInfoList.get(i).get("lowTem")){
116                  temp=(Integer)weatherInfoList.get(i).get("lowTem");
117              }
118          }
119          return temp;
120      }
121      public void getWeatherInfo(){                    //获取天气信息,解析字符串
122          weatherObject=MainActivity.weatherObject;
123          for(int i=0;i<5;i++){
             //获取未来五天的天气信息,并把每天的天气信息保存到一个Map对象中
124              Map<String,Object>item=new HashMap<String,Object>();
125              String dateInfo=weatherObject.getProperty(7+5*i).
                     toString();                          //日期信息和天气
126              String date=dateInfo.split(" ")[0];      //获取日期信息
127              String weather=dateInfo.split(" ")[1];   //获取天气信息
128              String temperInfo=weatherObject.getProperty(7+5*i+1)
129                  .toString();                         //温度信息
130              String lowTem=temperInfo.split("/")[0].substring(0,
131                  temperInfo.split("/")[0].lastIndexOf("℃")); //最低温度
132              String topTem=temperInfo.split("/")[1].substring(0,
```

```
133                 temperInfo.split("/")[1].lastIndexOf("℃"));   //最高温度
134             String icon1Str=weatherObject.getProperty(7+5*i+3)
135                 .toString();                                   //天气图标1
136             String icon2Str=weatherObject.getProperty(7+5*i+4)
137                 .toString();                                   //天气图标2
138             item.put("date",date);                             //日期
139             item.put("weather",weather);              //天气：晴、阴、多云等
140             item.put("topTem",Integer.parseInt(topTem));
141             item.put("lowTem",Integer.parseInt(lowTem));
142             item.put("icon1",Util.nameToImageId(icon1Str,"c_"));
143             item.put("icon2",Util.nameToImageId(icon2Str,"c_"));
144             weatherInfoList.add(item);
145         }
146     }
147 }
```

选择城市页面布局文件：16\WeatherForecast\res\layout\choose_city.xml

```
1  <LinearLayout xmlns:android="http://schemas.android.com/apk/res/android"
2      android:layout_width="match_parent"
3      android:layout_height="match_parent"
4      android:background="@drawable/body_bg"
5      android:orientation="vertical">
6      <ImageView
7          android:layout_width="match_parent"
8          android:layout_height="wrap_content"
9          android:src="@drawable/title_bg"
10         android:contentDescription="@string/imageInfo"/>
11     <TextView
12         android:text="@string/cityTitle"
13         android:layout_width="match_parent"
14         android:layout_height="wrap_content"
15         android:gravity="center"
16         android:textSize="20sp"
17         android:padding="10dp"/>
18     <GridView
19         android:id="@+id/provinceList"
20         android:layout_height="wrap_content"
21         android:layout_width="wrap_content"
22         android:numColumns="4"
23         android:layout_margin="10dp"
24         android:verticalSpacing="5dp"
25         android:horizontalSpacing="10dp"/>
26 </LinearLayout>
```

选择城市主程序：16\WeatherForecast\src\iet\jxufe\cn\android\weatherforecast\ChooseCityActivity.java

```java
1   public class ChooseCityActivity extends Activity {        //选择省份、城市页面
2       private GridView provinceList;                         //省份列表
3       private List<String>cities;                            //某个省中所有城市的集合
4       private List<String>provinces;                         //所有省份的集合
5       private String province,city;                          //获取城市和省份信息
6       protected void onCreate(Bundle savedInstanceState){
7           super.onCreate(savedInstanceState);
8           requestWindowFeature(Window.FEATURE_NO_TITLE);     //去除标题栏
9           getWindow().setFlags(WindowManager.LayoutParams.FLAG_FULLSCREEN,
10              WindowManager.LayoutParams.FLAG_FULLSCREEN);   //全屏显示
11          this.setContentView(R.layout.choose_city);
12          provinceList=(GridView)findViewById(R.id.provinceList);
13          getProvinces();                                    //获取省份信息
14          provinceList.setOnItemClickListener(new MyItemClickListener());
            //为GridView控件添加单击事件
15      }
16      private Handler mHandler=new Handler(){                //创建Handler对象
17          public void handleMessage(Message msg){            //处理消息的方法
18              if (msg.what ==0x11){                          //判断信息类型，0x11表示获取省份结束
19                  if (provinces !=null){
                    //如果省份集合不为空,将省份填充GridView控件
20                      ArrayAdapter<String>provinceAdapter=new
                            ArrayAdapter<String>(ChooseCityActivity.this,
21                          R.layout.province_item,provinces);
22                      provinceList.setAdapter(provinceAdapter);
23                  }
24              }else if(msg.what==0x12){                      //0x12表示获取城市结束
25                  if(cities!=null){
26                      createDialog();                        //弹出对话框,让用户选择城市
27                  }
28              }
29          }
30      };
31      public void getProvinces(){    //获取省份信息,启动线程,向Web Service发送请求
32          new Thread(){                                      //创建线程
33              public void run(){
34                  provinces=WebServiceUtil.getProvinces();
                    //调用Web Service辅助类的方法
35                  mHandler.sendEmptyMessage(0x11);           //得到信息后发送消息
36              }
37          }.start();                                         //启动线程
38      }
```

```
39  private class MyItemClickListener implements OnItemClickListener{
        //省份项的单击事件处理
40      public void onItemClick(AdapterView<?>parent,View view,int position,
41          long id){
42          province=provinces.get(position);        //获取单击的省份
43          getCitiesByProvince(province);           //根据省份查询城市
44      }
45  }
46  public void createDialog(){
47      AlertDialog.Builder builder=new Builder(this);
48      builder.setTitle("请选择城市");
49      View citySelect=getLayoutInflater().inflate(R.layout.city_
            select,null);
50      final TextView cityText=(TextView)citySelect.findViewById
            (R.id.selectedCity);                     //显示当前选中的城市
51      city=cities.get(0);                          //默认显示第一个城市
52      cityText.setText(Html.fromHtml("默认的城市为:<font color=blue><b>"
            +city+"</b></font>"));
53      GridView cityView= (GridView)citySelect.findViewById(R.id.cityList);
        //显示城市的列表
54      ArrayAdapter<String>cityAdapter=new ArrayAdapter<String>(this,
            R.layout.province_item,cities);
55      cityView.setAdapter(cityAdapter);
56      cityView.setOnItemClickListener(new OnItemClickListener(){
57          public void onItemClick(AdapterView<?>parent,View view,
58              int position,long id){
59              city=cities.get(position);
60              cityText.setText(Html.fromHtml("你选择的城市为:<font
                    color=blue><b>"+city+"</b></font>"));
61          }
62      });
63      builder.setView(citySelect);    //自定义对话框,对话框内容为指定的View
64      builder.setPositiveButton("确定",new OnClickListener(){
65          public void onClick(DialogInterface dialog,int which){
66              Intent intent=getIntent();           //获取 Intent 对象,传递数据
67              intent.putExtra("city",city);        //保存当前选中的城市
68              intent.putExtra("province",province);//保存当前选中的省份信息
69              setResult(0,intent);                 //将数据返回给 MainActivity
70              ChooseCityActivity.this.finish();    //结束当前的 Activity
71          }
72      });
73      builder.setCancelable(false);                //对话框不可取消
74      builder.setNegativeButton("取消",null);       //"取消"按钮
75      builder.create().show();                     //创建并显示对话框
```

```
76        }
77    public void getCitiesByProvince(final String province){
          //根据省份获取该省份的所有城市
78        new Thread(){                                    //创建线程
79            public void run(){
80                cities=WebServiceUtil.getCitiesByProvince(province);
                  //调用 Web Service 辅助类的方法
81                mHandler.sendEmptyMessage(0x12);
82            }
83        }.start();                                       //启动线程
84    }
85  }
```

省份项的布局文件：16\WeatherForecast\res\layout\province_item.xml

```
1   <TextView xmlns:android="http://schemas.android.com/apk/res/android"
2       android:textSize="18sp"
3       android:layout_width="match_parent"
4       android:layout_height="match_parent"
5       android:gravity="center"
6       android:singleLine="true"
7       android:ellipsize="end"
8       android:background="@drawable/btn_bg"/>
```

自定义对话框对应的布局文件：16\WeatherForecast\res\layout\city_select.xml

```
1   <LinearLayout xmlns:android="http://schemas.android.com/apk/res/android"
2       android:layout_width="match_parent"
3       android:layout_height="match_parent"
4       android:orientation="vertical">
5       <TextView
6           android:id="@+id/selectedCity"
7           android:layout_width="match_parent"
8           android:layout_height="wrap_content"
9           android:gravity="center"
10          android:textSize="20sp"
11          android:padding="10dp"/>
12      <GridView
13          android:id="@+id/cityList"
14          android:layout_height="wrap_content"
15          android:layout_width="wrap_content"
16          android:numColumns="3"
17          android:layout_margin="10dp"
18          android:verticalSpacing="5dp"
```

```
19        android:horizontalSpacing="10dp"/>
20   </LinearLayout>
```

按钮背景：16\WeatherForecast\res\drawable\btn_bg.xml

```
1  <?xml version="1.0" encoding="utf-8"?>
2  <selector xmlns:android="http://schemas.android.com/apk/res/android">
3      <item android:state_selected="true"
           android:drawable="@drawable/shape_pressed"></item>
4      <item android:state_pressed="true"
           android:drawable="@drawable/shape_pressed"></item>
5      <item android:drawable="@drawable/shape_unpressed"></item>
6  </selector>
```

天气带边框信息：16\WeatherForecast\res\drawable\weather_boder.xml

```
1  <?xml version="1.0" encoding="utf-8"?>
2  <shape xmlns:android="http://schemas.android.com/apk/res/android"
3      android:shape="rectangle">
4      <corners android:radius="10dp"/>
5      <solid android:color="#00000000"/>
6      <stroke
7          android:width="2dp"
8          android:color="#770000bb"
9          android:dashWidth="2dp"
10         android:dashGap="3dp"/>
11 </shape>
```

自定义按下时图片：16\WeatherForecast\res\drawable\shape_pressed.xml

```
1  <?xml version="1.0" encoding="utf-8"?>
2  <shape xmlns:android="http://schemas.android.com/apk/res/android"
3      android:shape="rectangle">
4      <corners android:radius="5dp"/>
5      <solid android:color="#aa00ffff"/>
6  </shape>
```

自定义正常时图片：16\WeatherForecast\res\drawable\shape_unpressed.xml

```
1  <?xml version="1.0" encoding="utf-8"?>
2  <shape xmlns:android="http://schemas.android.com/apk/res/android"
3      android:shape="rectangle">
4      <corners android:radius="5dp"/>
5      <solid android:color="#3300cccc"/>
6  </shape>
```

工具类：16\WeatherForecast\src\iet\jxufe\cn\android\weatherforecast\Util.java

```java
1   public class Util {
2       public static int nameToImageId(String icon,String pre){
            //根据文件名获取对应图片 ID
3           String iconName=pre+ icon.substring(0, icon.lastIndexOf("."));
4           try {
5               Field field=R.drawable.class.getField(iconName);
                //根据名称获取成员变量
6               int resId=field.getInt(R.drawable.class);    //获取成员变量对应的值
7               return resId;
8           } catch(Exception e){
9               e.printStackTrace();
10          }
11          return R.drawable.b_0;                           //默认返回晴天的图标
12      }
13      public static String intToWeek(int dayOfWeek){  //将序号转换成对应的星期
14          switch(dayOfWeek){
15          case Calendar.SUNDAY:
16              return "星期天";
17          case Calendar.MONDAY:
18              return "星期一";
19          case Calendar.TUESDAY:
20              return "星期二";
21          case Calendar.WEDNESDAY:
22              return "星期三";
23          case Calendar.THURSDAY:
24              return "星期四";
25          case Calendar.FRIDAY:
26              return "星期五";
27          case Calendar.SATURDAY:
28              return "星期六";
29          }
30          return "不合法数字";
31      }
32  }
```

16.3 代码分析

16.3.1 调用 Web Service

手机的计算能力、存储能力都比较有限，通常是作为移动终端来使用，具体的数据处理是交给网络服务器进行，手机的主要优势在于携带方便，可随时随地访问网络、获取数

第16章 天气预报——Web Service 的调用

据。访问网络数据又可分为访问自己开发的服务器端程序和访问第三方提供的服务，本案例就是通过 Web Service 访问第三方服务，充当 Web Service 的客户端。

Web Service 是一种基于 SOAP（Simple Object Access Protocol，简单对象访问协议）协议的远程调用标准，主要包括 SOAP、WSDL（Web Service Description Language，Web Service 描述语言）、UDDI（Universal Description Discovery and Integration，统一描述、发现和整合协议）3 个要素。其中，SOAP 用于传递信息的格式，WSDL 用于描述如何访问具体的接口，UDDI 用于管理、分发、查询 Web Service。通过 Web Service 可以将不同操作系统平台、不同语言、不同技术整合到一起。

在 Android SDK 中并没有提供调用 Web Service 的库，因此需要使用第三方的 SDK 来调用 Web Service。比较常用的有 ksoap2，用户可以通过 http://ksoap2-android.googlecode.com/svn/m2-repo/com/google/code/ksoap2-android/ksoap2-android-assembly/3.1.0/ksoap2-android-assembly-3.1.0-jar-with-dependencies.jar 进行下载。然后将下载得到的 jar 包添加到 Android 项目的 libs 目录下，接下来即可使用相关 API 调用 Web Service 所暴露的操作。

使用 ksoap2-android 调用 Web Service 操作的步骤如下：

（1）创建 HttpTransportSE 对象，该对象用于发送请求，调用 Web Service 需要传递想要访问的服务地址。

（2）创建 SoapSerializationEnvelope 对象，该对象代表 SOAP 消息封装包，用户请求的 SOAP 以及服务器响应生成的 SOAP 都可以通过该对象设置和得到，在创建该对象时需要传递当前使用的 SOAP 的版本号，不同的版本会有所区别。

（3）创建 SoapObject 对象，在创建该对象时需要传递所需要调用的 Web Service 的命名空间以及访问的方法名，其中，命名空间需要和 WSDL（Web Service Description Language，Web Service 描述语言）文件中描述的一致。

（4）如果调用方法时需要传递参数，则调用 SoapObject 对象的 addProperty（String name，Object value）方法来设置参数。其中，name 表示参数名称，value 表示参数对应的值，可多次调用该方法设置多个参数。

（5）调用 SoapSerializationEnvelope 对象的 setOutputSoapObject（）方法，或者直接对 bodyOut 属性赋值，将前面所创建的 SoapObject 对象作为请求体。

（6）调用 HttpTransportSE 对象的 call（）方法，发送请求，该方法需要传递两个参数，第一个参数是命名空间＋需要访问的方法名，第二个参数为 SoapSerializationEnvelope 对象。

（7）调用完成后，判断是否有响应，即 SoapSerializationEnvelope 对象的 getResponse（）方法的结果是否为空，如果不为空，则根据 SoapSerializationEnvelope 对象的 bodyIn 属性获取 SoapObject 对象，该对象代表了 Web Service 的返回信息，然后解析该 SoapObject 对象，即可获取 Web Service 的返回值。

下面以获取天气信息为例进行介绍，首先找到第三方提供天气预报的 Web Service 的地址，在此为 http://webservice.webxml.com.cn/WebServices/WeatherWS.asmx，该页面提供了相应的方法，根据这些方法可以获取相应的数据，其中有一个方法是 getWeather（），该方法表示根据城市来查询天气，需要传递一个参数城市名或者注册的用户 ID。打开该方法的链接，可知对应的命名空间为 http://WebXml.com.cn/，两个参数的名称分别为 theCityCode 和 theUserID，获取方法结果的属性名为 getWeatherResult。其详细代码见 WebServiceUtil.java 的第 51~71 行。

16.3.2 用 SharedPreference 保存用户信息

通常，用户在使用 Android 应用时都会根据自己的爱好进行简单的设置，例如改变背景颜色、记录用户名和密码、登录状态等，为了使用户下次打开应用时不需要重复设置，Android 应用应该保存这些设置信息。通常这些信息相对来说比较简单，在 Android 中可使用 SharedPreference 保存这些信息。SharedPreference 是一个轻量级的存储方式。

应用程序使用 SharedPreferences 可以快速、高效地以键值对的形式保存数据，非常类似于 Bundle。信息以 XML 文件的形式存储在 Android 设备上。SharedPreferences 本身是一个接口，不能直接实例化，只能通过 Context 提供的 getSharedPreferences(String name, int mode) 方法获取 SharedPreferences 实例，第一个参数表示保存信息的文件名，不需要后缀；第二个参数表示 SharedPreferences 的访问权限，包括只能被本应用程序读、写，能被其他应用程序读、能被其他应用程序写。

在得到 SharedPreference 对象后，即可调用它的一系列 getXXX() 方法获取相应关键字对应的值，例如 getInt()、getString() 等。该方法需要传递两个参数，第一个参数为关键字，即保存时使用的关键字；第二个参数为默认值，即没有该关键字时返回的值。在 SharedPreference 中只能保存 int、float、long、boolean、String、Set<String> 等少数比较简单的类型数值。代码见 MainActivity.java 的第 15~17 行。

SharedPreferences 接口本身只提供了读取数据的功能并没有提供写入数据的功能，如果需要实现写入功能，则需通过 SharedPreferences 的内部接口 Editor 来实现。SharedPreferences 调用 edit() 方法即可获取它对应的 Editor 对象，然后调用该对象的一系列的 putXXX() 方法保存数据，最后调用 Editor 对象的 commit() 方法提交数据。代码见 MainActivity.java 的第 70~73 行。

注意：当程序所读取的 SharedPreferences 文件不存在时，程序也会返回默认值，并不会抛出异常。SharedPreferences 数据总是保存在 /data/data/<package_name>/shared_prefs 目录下，并且 SharedPreferences 数据总是以 XML 格式保存。例如本例切换到 DDMS 视图，打开 File Explore 面板，展开文件浏览树，将会发现在 /data/data/ iet.jxufe.cn.android.weatherforecast/shared_prefs 目录下生成了上面定义的 weather.xml 文件，如图 16-7 所示。

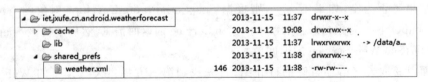

图 16-7 SharedPreferences 数据保存的路径

将该文件下载到计算机上打开，文件内容如下。

```
1  <?xml version='1.0' encoding='utf-8' standalone='yes'?>
2  <map>
3    <string name="province">山东</string>
4    <string name="city">济南</string>
5  </map>
```

16.3.3 按两次返回键退出应用程序

在 Android 应用中，为了避免用户误操作退出应用程序，通常会在用户单击"退出"按钮或按手机上的返回键时给用户一些提示。例如弹出对话框询问用户是否确定要退出，或者提示用户再次单击退出应用。在本案例中使用了后者，连续单击两次"退出"按钮或按两次返回键时退出程序，第一次单击时弹出如图 16-8 所示的信息。

其主要思路是定义一个全局变量 count 用于记录用户单击"退出"按钮的次数，默认为 1，然后为"退出"按钮添加事件处理，首先判断 count 的值是否小于 2，如果小于 2 则通过 Toast 弹出提示信息，并启动一个线程用于计时，3s 后 count 的值恢复为 1，如果在 3s 内单击第二次则退出。如果不小于 2，调用 Activity 的 finish() 方法结束当前 Activity。代码见 MainActivity 的第 78～95 行。

对返回键的处理类似，只是需要捕捉用户的按键操作，重写 Activity 的按键事件回调方法

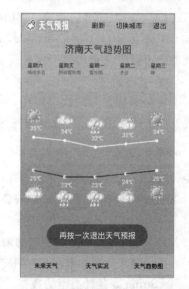

图 16-8 单击退出键时弹出的提示信息

onKeyDown(int keyCode，KeyEvent event)，判断用户按下的是否是返回键，如果是，处理调用上面"退出"按钮的执行方法。代码见 MainActivity 的第 103～109 行。

16.4 知识扩展

在 Android 中，Activity 之间的数据传递方向有两种，一种是从 AActivity 启动 BActivity，并传递数据给 BActivity；另外一种是从 AActivity 启动 BActivity，当 BActivity 结束的时候，将数据从 BActivity 中返回给 AActivity。前一种比较简单，在前面的实例中也多次用到（例如南昌景点介绍实例），只需在 AActivity 中通过 startActivity(Intent intent) 启动 BActivity，并将数据放在 Intent 中即可。相对来说第二种比较复杂，首先在 AActivity 中通过 startActivityForResult((Intent intent，int requestCode) 方法启动 BActivity，其中，requestCode 表示请求码，同时还需要在 AActivity 中重写其 onActivityResult(int requestCode，int resultCode，Intent intent) 方法，其中 requestCode 代表请求码，resultCode 代表返回的结果码。此外，在 BActivity 结束之前，调用 BActivity 的 setResult(int resultCode，Intent intent) 方法，resulutCode 表示结果码，将需

要返回的结果写入到 Intent 中，传递给 AActivity。

整个执行过程为 AActivity 调用 startActivityForResult（Intent intent，int requestCode）方法启动 BActivity 之后执行 BActivity 的相应方法，并将执行结果通过 setResult(int resultCode，Intent intent)方法写入 Intent，当 BActivity 执行结束后，会调用 AActivity 的 onActivityResult(int requestCode，int resultCode，Intent intent)，在该方法中判断请求码和结果码是否符合要求，从而获取 Intent 中的数据。

请求码和结果码的作用：因为在一个 Activity 中可能存在多个控件，每个控件都有可能添加相应的事件处理，调用 startActivityForResult()方法，从而有可能打开多个不同的 Activity 处理不同的业务。但这些 Activity 关闭之后，都会调用原来 Activity 的 onActivityResult(int requestCode，int resultCode，Intent intent)方法。通过请求码，我们就能知道该方法是由哪个控件所触发的，通过结果码，我们就能知道返回的数据来自于哪个 Activity。

不管是哪种方向的数据传递，数据都保存在 Intent 中，那么 Intent 是如何保存数据的呢？Intent 提供了多个重载方法来存放数据，主要格式如下：

putExtras(String name,XXX data)

其中，XXX 表示数据类型，向 Intent 中放入 XXX 类型的数据，例如 int、long、String 等，同时 Intent 中也提供了相应的 getXXXExtras(String name)方法来获取数据。此外，还提供了一个 putExtras(Bundle data)方法，该方法可用于存放一个数据包，Bundle 类似于 Java 中的 Map 对象，存放的是键值对的集合，可把多个相关数据放入同一个 Bundle 中，Bundle 提供了一系列存入数据的方法，方法格式为 putXXX(String key，XXX data)，向 Bundle 中放入 int、long、String 等各种类型的数据。为了取出 Bundle 数据携带包中的数据，Bundle 还提供了相应的 getXXX(String key)方法，从 Bundle 中取出各种类型的数据。

16.5 思考与练习

模仿调用天气的 Web Service，调用 Web Service 实现国内手机号码归属地查询功能。（Web Service 地址：http://webservice.webxml.com.cn/WebServices/MobileCodeWS.asmx）

第17章

音乐播放器

17.1 案例概述

本案例主要实现音乐播放器功能,音乐播放是一种比较耗时的操作,通常将其放在后台执行,而主线程可以继续做其他的事情,例如一边浏览网页一边听音乐,即播放音乐的应用退出后仍可以继续播放音乐。前、后台的交互主要是通过广播来完成的,当前台需要对后台的音乐播放进行控制时,例如第一首、上一首、播放/暂停、下一首、最后一首,发送一条广播,后台服务接收到广播后对其进行相应处理。当音乐播放完成后,根据用户设置的播放类型(如列表循环、单曲循环、随机播放等)进行后期处理。除此之外,本案例还实现了将某一首音乐设置为手机铃声等功能,程序运行效果如图17-1至图17-9所示。

图 17-1 程序运行主界面

图 17-2 选中音乐长按弹出上下文菜单

图 17-3 音乐按艺术家分类

图 17-4 音乐按专辑分类

图 17-5 播放列表

图 17-6 播放列表中对音乐的操作

Android 编程经典案例解析

图 17-7　音乐播放界面　　　图 17-8　选择播放形式界面　　　图 17-9　音乐状态通知

17.2　关键代码

主界面布局文件：17\MusicPlayer\res\layout\activity_main.xml

```
1    <TabHost xmlns:android="http://schemas.android.com/apk/res/android"
2        android:id="@+id/mTabHost"
3        android:layout_width="match_parent"
4        android:layout_height="match_parent" >
5        <LinearLayout
6            android:layout_width="match_parent"
7            android:layout_height="match_parent"
8            android:orientation="vertical" >
9            <TabWidget
10               android:id="@android:id/tabs"
11               android:layout_width="match_parent"
12               android:layout_height="wrap_content"/>
13           <FrameLayout
14               android:id="@android:id/tabcontent"
15               android:layout_width="match_parent"
16               android:layout_height="0dp"
17               android:layout_weight="1"
18               android:background="@drawable/content_bg">
19               <FrameLayout
20                   android:id="@+id/realContent"
21                   android:layout_width="match_parent"
22                   android:layout_height="match_parent" />
23           </FrameLayout>
24       </LinearLayout>
```

```
25    </TabHost>
```

单个选项对应的布局文件：17\MusicPlayer\res\layout\tab.xml

```
1   <LinearLayout xmlns:android="http://schemas.android.com/apk/res/android"
2       xmlns:tools="http://schemas.android.com/tools"
3       android:layout_width="match_parent"
4       android:layout_height="match_parent"
5       android:gravity="center_horizontal"
6       android:padding="5dp"
7       android:background="@drawable/bg"
8       android:orientation="vertical">
9       <ImageView
10          android:id="@+id/icon"
11          android:layout_width="30dp"
12          android:layout_height="30dp"
13          android:contentDescription="@string/imageInfo" />
14      <TextView
15          android:id="@+id/title"
16          android:textColor="#ffffff"
17          android:textStyle="bold"
18          android:textSize="12sp"
19          android:layout_width="wrap_content"
20          android:layout_height="wrap_content" />
21  </LinearLayout>
```

程序主界面：17\MusicPlayer\src\iet\jxufe\cn\android\musicplayer\MainActivity.java

```
1   public class MainActivity extends Activity {              //主界面
2       private TabHost mTabHost;                             //选项卡
3       private String[]titles=new String[]{"艺术家","音乐","专辑",
            "播放列表"};                                       //选项标题
4       private String[]tags=new String[]{"artist","music","album",
            "playlist"};                                      //选项标记
5       private int[]icons=new int[]{R.drawable.music,R.drawable.artist,
6           R.drawable.album,R.drawable.playlist};            //选项图标
7       private MyOpenHelper mHelper;                         //数据库辅助类
8       private SQLiteDatabase mDatabase;                     //数据库类
9       protected void onCreate(Bundle savedInstanceState){
10          super.onCreate(savedInstanceState);
11          requestWindowFeature(Window.FEATURE_NO_TITLE);    //去除标题栏
12          getWindow().setFlags(WindowManager.LayoutParams.FLAG_FULLSCREEN,
13              WindowManager.LayoutParams.FLAG_FULLSCREEN);//全屏显示
14          setContentView(R.layout.activity_main);
```

```
15          initData();                                           //初始化数据
16          mTabHost=(TabHost)findViewById(R.id.mTabHost);
17          mTabHost.setup();
18          for(int i=0;i<titles.length;i++){                     //循环添加选项卡
19              TabSpec tabSpec=mTabHost.newTabSpec(tags[i]);
                //创建一个选项,并指定其标记
20              View view=getLayoutInflater().inflate(R.layout.tab,null);
                //将布局文件转换为 View 对象
21              TextView titleView=(TextView)view.findViewById(R.id.title);
22              ImageView iconView=(ImageView)view.findViewById(R.id.icon);
23              titleView.setText(titles[i]);                     //设置标题
24              iconView.setImageResource(icons[i]);              //设置图标
25              tabSpec.setIndicator(view);                       //为选项设置标题和图标
26              tabSpec.setContent(R.id.realContent);             //为每个选项设置内容
27              mTabHost.addTab(tabSpec);                         //将选项添加到选项卡中
28          }
29          mTabHost.setOnTabChangedListener(new MyTabChangedListener());
            //添加选项改变事件处理
30          mTabHost.setCurrentTab(1);                            //默认显示第二个
31      }
32      public void initData(){
33          mHelper=new MyOpenHelper(this,"music",null,1);        //得到数据库辅助类
34          mDatabase=mHelper.getWritableDatabase();              //获取数据库
35          Constants.playlist=MusicUtils.getDataFromDB(mDatabase);
            //初始化播放列表
36      }
37      private class MyTabChangedListener implements OnTabChangeListener {
            //选项改变事件监听器
38          public void onTabChanged(String tabTag){
39              FragmentTransaction fragmentTransaction=getFragmentManager()
40                  .beginTransaction();                          //开始事务
41              if(tabTag.equalsIgnoreCase("music")){             //切换到音乐列表
42                  MusicListFragment musicListFragment=new MusicListFragment();
43                  musicListFragment.setMusicList(null);
44                  fragmentTransaction.replace(R.id.realContent,
45                      musicListFragment);
46              }else if(tabTag.equalsIgnoreCase("artist")){      //切换到按艺术家分类
47                  fragmentTransaction.replace(R.id.realContent,
48                      new ArtistListFragment());
49              }else if(tabTag.equalsIgnoreCase("album")){       //切换到按专辑分类
50                  fragmentTransaction.replace(R.id.realContent,
51                      new AlbumListFragment());
52              }else if(tabTag.equalsIgnoreCase("playlist")){
53                  fragmentTransaction.replace(R.id.realContent,
```

```
54                    new PlayListFragment());
55            }
56            fragmentTransaction.commit();                //提交事务
57        }
58    }
59    protected void onDestroy(){  //关闭的时候将播放列表中的数据保存到数据库中
60        mDatabase.execSQL("delete from music_tb");        //删除已有的所有数据
61        for(int i=0;i<Constants.playlist.size();i++){//循环遍历播放列表中的音乐
62            Music music=Constants.playlist.get(i);       //获取音乐
63            mDatabase.execSQL("insert into music_tb (title,artist,album,
                album_id,time,url)values(?,?,?,?,?,?)",new String[]{
                music.getTitle(),music.getSinger(),music.getAlbum(),
64              music.getAlbum_id()+"",music.getTime()+"",music.getUrl()});
                //将音乐信息保存到数据库
65        }
66        super.onDestroy();
67    }
68 }
```

音乐封装类：17\MusicPlayer\src\iet\jxufe\cn\android\musicplayer\Music.java

```
1  public class Music implements Serializable {
2      private static final long serialVersionUID=1;
3      private String title;              //歌曲文件标题
4      private String singer;             //歌曲演唱者
5      private String album;              //歌曲专辑
6      private int album_id;              //专辑编号
7      private String url;                //歌曲文件路径
8      private long size;                 //歌曲文件大小
9      private int time;                  //歌曲文件时长,单位为毫秒
10     private String name;               //歌曲文件名,包含后缀
11     public int getAlbum_id(){
12         return album_id;
13     }
14     public void setAlbum_id(int album_id){
15         this.album_id=album_id;
16     }
17     public String getTitle(){
18         return title;
19     }
20     public void setTitle(String title){
21         this.title=title;
22     }
23     public String getSinger(){
```

```
24         return singer;
25     }
26     public void setSinger(String singer){
27         this.singer=singer;
28     }
29     public String getAlbum(){
30         return album;
31     }
32     public void setAlbum(String album){
33         this.album=album;
34     }
35     public String getUrl(){
36         return url;
37     }
38     public void setUrl(String url){
39         this.url=url;
40     }
41     public long getSize(){
42         return size;
43     }
44     public void setSize(long size){
45         this.size=size;
46     }
47     public int getTime(){
48         return time;
49     }
50     public void setTime(int time){
51         this.time=time;
52     }
53     public String getName(){
54         return name;
55     }
56     public void setName(String name){
57         this.name=name;
58     }
59     public String toString(){          //显示歌曲名和演唱者
60         return "Music [title="+title+",singer="+singer+"]";
61     }
62 }
```

获取音乐辅助类：17\MusicPlayer\src\iet\jxufe\cn\android\musicplayer\MusicUtils.java

```
1  public class MusicUtils {
2      public static List<Music>getMusicData(Context context){
```

```
3            ContentResolver mResolver=context.getContentResolver();
             //获取内容解析器
4            if(mResolver!=null){                              //获取所有歌曲
5                //第一个参数表示系统中音乐提供者的URI
                 //第二个参数表示需要获取的列的信息
                 //第三个参数表示查询条件
                 //第四个参数表示条件中的占位符赋值
                 //第五个参数表示查询结果的排序方式
                 Cursor cursor=mResolver.query(
                     MediaStore.Audio.Media.EXTERNAL_CONTENT_URI,null,null,
                     null,MediaStore.Audio.Media.DEFAULT_SORT_ORDER);
6                return cursorToList(cursor,context);
7            }
8            return null;
9        }
10       public static List<Music>cursorToList(Cursor cursor,Context context){
11           if(cursor==null||cursor.getCount()==0){
12               return null;
13           }
14           List<Music>musicList=new ArrayList<Music>();    //用于存放音乐信息的集合
15           while(cursor.moveToNext()){
16               Music m=new Music();                        //创建音乐对象
17               String title=cursor.getString(cursor
18                   .getColumnIndex(MediaStore.Audio.Media.TITLE));  //获取音乐标题
19               String artist=cursor.getString(cursor
20                   .getColumnIndex(MediaStore.Audio.Media.ARTIST));//获取音乐艺术家
21               if("<unknown>".equals(artist)){
22                   artist="未知艺术家";
23               }
24               String album=cursor.getString(cursor.getColumnIndex
                     (MediaStore.Audio.Media.ALBUM));       //获取音乐专辑
25               int album_id=cursor.getInt(cursor.getColumnIndex
                     (MediaStore.Audio.Media.ALBUM_ID));
26               long size=cursor.getLong(cursor.getColumnIndex
                     (MediaStore.Audio.Media.SIZE));        //获取音乐大小
27               int time=cursor.getInt(cursor.getColumnIndex
                     (MediaStore.Audio.Media.DURATION));
                 //获取音乐持续的时间,单位为毫秒
28               String url=cursor.getString(cursor.getColumnIndex
                     (MediaStore.Audio.Media.DATA));        //获取音乐的保存路径
29               String name=cursor.getString(cursor
                     .getColumnIndex(MediaStore.Audio.Media.DISPLAY_NAME));
                 //获取音乐名,包含后缀
30               String sub=name.substring(name.lastIndexOf(".")+1);
```

```java
31            //获取文件的扩展名
32            if(sub.equals("mp3")&& time >50000){ //以 mp3 结尾并且长度大于 5s
33                m.setTitle(title);              //歌曲标题
34                m.setSinger(artist);            //歌曲的演唱者
35                m.setAlbum(album);              //歌曲所属专辑
36                m.setAlbum_id(album_id);        //歌曲所属专辑的编号
37                m.setSize(size);                //歌曲的大小
38                m.setTime(time);                //歌曲的时长
39                m.setUrl(url);                  //歌曲存放的路径
40                m.setName(name);                //歌曲名,包含后缀
41                musicList.add(m);               //将歌曲添加到集合中
42            }
43        }
44        cursor.close();                         //关闭游标
45        return musicList;
46    }
47    public static String timeToString(int time){
          //时间格式转换,将毫秒转换成分秒的形式
48        int temp=time/1000;                     //将毫秒转换成秒
49        int minute=temp/60;                     //计算一共有多少分
50        int second=temp%60;                     //除了这些分,还剩多少秒
51        return String.format("%02d:%02d",minute,second);   //以分秒的形式显示
52    }
53    public static List<Music>getDataFromDB(SQLiteDatabase db){
54        List<Music>musics=new ArrayList<Music>();
55        Cursor cursor=db.rawQuery("select * from music_tb",null);
56        if(cursor==null||cursor.getCount()==0){
57            return musics;
58        }
59        while(cursor.moveToNext()){
60            Music music=new Music();
61            music.setTitle(cursor.getString(cursor.getColumnIndex
                (("title"))));
62            music.setSinger(cursor.getString(cursor.getColumnIndex
                ("artist")));
63            music.setAlbum(cursor.getString(cursor.getColumnIndex
                ("album")));
64            music.setAlbum_id(cursor.getInt(cursor.getColumnIndex
                ("album_id")));
65            music.setUrl(cursor.getString(cursor.getColumnIndex("url")));
66            music.setTime(cursor.getInt(cursor.getColumnIndex("time")));
67            musics.add(music);
68        }
69        return musics;
```

```
70      }
71      public static Bitmap getAlbumPic(Context context,Music music){
72          ContentResolver mResolver=context.getContentResolver();
            //获取内容解析器
73          Uri uri=ContentUris.withAppendedId(Constants.ALBUM_URL,
                music.getAlbum_id());
74          try {
75              InputStream inputStream=mResolver.openInputStream(uri);
76              return BitmapFactory.decodeStream(inputStream);
77          } catch(FileNotFoundException ex){        //如果不存在则抛出异常
78              try {
79                  ParcelFileDescriptor pfd=mResolver.openFileDescriptor
                        (uri,"r");
80                  if(pfd !=null){
81                      FileDescriptor fd=pfd.getFileDescriptor();
82                      Bitmap bitmap=BitmapFactory.decodeFileDescriptor(fd);
83                      return bitmap;
84                  }
85              } catch(Exception e){
86                  return null;
87              }
88          return null;
89          }
90      }
91  }
```

数据库辅助类：17\MusicPlayer\src\iet\jxufe\cn\android\musicplayer\MyOpenHelper.java

```
1   public class MyOpenHelper extends SQLiteOpenHelper {
2       public String createTableSQL="create table if not exists music_tb" +
            "(_id integer primary key autoincrement,title,artist,album,
            album_id,time,url)";
3       public MyOpenHelper(Context context,String name,CursorFactory
            factory,int version){
4           super(context,name,factory,version);
5       }
        //数据库创建后回调该方法,执行建表操作和插入初始化数据的操作
7       public void onCreate(SQLiteDatabase db){
8           db.execSQL(createTableSQL);
9       }
10      //数据库版本更新时回调该方法
11      public void onUpgrade(SQLiteDatabase db,int oldVersion,int newVersion){
12          System.out.println("版本变化:"+oldVersion+"-------->"+newVersion);
13      }
```

14 }

音乐列表 Fragment：17\MusicPlayer\src\iet\jxufe\cn\android\musicplayer\MusicListFragment.java

```java
1   public class MusicListFragment extends ListFragment {
        //选项卡默认显示所有音乐信息,所以在此启动后台音乐播放服务
2       public List<Music>musicList;                              //要显示的音乐的集合
3       public void onCreate(Bundle savedInstanceState){
4           super.onCreate(savedInstanceState);
5           musicList=getMusicList();                             //获取音乐
6           if(musicList==null||musicList.size()==0){             //如果音乐为空
7               musicList=MusicUtils.getMusicData(getActivity()); //获取所有音乐
8           }
9       }
10      public View onCreateView(LayoutInflater inflater,ViewGroup container,
11          Bundle savedInstanceState){
12          Constants.musiclist=musicList;
13          if(musicList !=null){
14              setListAdapter(new MusicAdapter());               //显示音乐列表
15          } else {
16              Toast.makeText(getActivity(),"存储卡中暂时没有音乐,请添加音乐{...",
17                  Toast.LENGTH_SHORT).show();                   //提示存储卡中没有音乐
18          }
19          Intent intent=new Intent(getActivity(),MusicService.class);
            //创建 Intent,启动指定的服务
20          getActivity().startService(intent);                   //启动服务
21          return super.onCreateView(inflater,container,savedInstanceState);
22      }
23      public void onStart(){
24          registerForContextMenu(getListView()); //为音乐列表注册上下文菜单
25          super.onStart();
26      }
27      private class MusicAdapter extends BaseAdapter {
28          public int getCount(){
29              return musicList.size();
30          }
31          public Object getItem(int position){
32              return musicList.get(position);
33          }
34          public long getItemId(int position){
35              return position;
36          }
37          public View getView(int position,View convertView,ViewGroup parent){
38              if(convertView==null){
```

```
39              convertView=LinearLayout.inflate(getActivity(),
40                  R.layout.music_item,null);   //将布局文件转换成 View 对象
41              }
42              ImageView icon=(ImageView)convertView.findViewById(R.id.icon);
                //显示图标的控件
43              TextView title=(TextView)convertView.findViewById(R.id.title);
                //显示歌曲名的控件
44              TextView artist=(TextView)convertView.findViewById(R.id.artist);
                //显示演唱者的控件
45              TextView time=(TextView)convertView.findViewById(R.id.time);
                //显示时间的控件
46              Bitmap bitmap=MusicUtils.getAlbumPic(getActivity(),
                    musicList.get(position));          //显示专辑图片的控件
47              if(bitmap!=null){//如果专辑图片不为空,则显示;如果为空,则显示默认图片
48                  icon.setImageBitmap(bitmap);
49              }else {
50                  icon.setImageResource(R.drawable.music); //显示默认的图片
51              }
52              title.setText(musicList.get(position).getTitle());
53              artist.setText(musicList.get(position).getSinger());
54              time.setText(MusicUtils.timeToString(musicList.get(position)
55                  .getTime()));
56              return convertView;
57          }
58      }
59      public List<Music>getMusicList(){
60          return musicList;
61      }
62      public void setMusicList(List<Music>musicList){
63         this.musicList=musicList;
64      }
65      public void onListItemClick(ListView l,View v,int position,long id){
            //音乐项被单击事件处理
66          Intent intent=new Intent(getActivity(),MusicPlayActivity.class);
67          intent.putExtra("listType",Constants.ALL_MUSIC);
            //音乐播放的列表类型:所有音乐的列表
68          intent.putExtra("music",musicList.get(position));   //当前的音乐
69          intent.putExtra("position",position);   //当前音乐对应的位置
70          startActivity(intent);
71          super.onListItemClick(l,v,position,id);
72      }
73      public void onCreateContextMenu(ContextMenu menu,View v,
74          ContextMenuInfo menuInfo){           //创建上下文菜单
75          getActivity().getMenuInflater().inflate(R.menu.musiclist_
```

```
                            context,menu);
76          super.onCreateContextMenu(menu,v,menuInfo);
77      }
78      public boolean onContextItemSelected(MenuItem item){
79          AdapterContextMenuInfo info=(AdapterContextMenuInfo)item
80              .getMenuInfo();                          //获取上下文菜单信息
81          switch(item.getItemId()){
82          case R.id.setToBell:                         //设置为手机铃声
83              setRing(musicList.get(info.position));
84              break;
85          case R.id.addToPlayList:                     //添加到收藏列表
86              Music music=musicList.get(info.position);
87              int i=0;
88              for(; i<Constants.playlist.size(); i++){
                    //循环遍历播放列表中是否已经存在该音乐
89                  if(Constants.playlist.get(i).getTitle()
90                      .equalsIgnoreCase(music.getTitle())){
91                      break;                           //如果存在则不需要添加,直接退出
92                  }
93              }
94              if(i==Constants.playlist.size()){
95                  Constants.playlist.add(music);
96              }
97              break;
98          default:
99              break;
100         }
101         return super.onContextItemSelected(item);
102     }
103     public void setRing(Music music){                //设置铃声
104         ContentValues values=new ContentValues();
105         values.put(MediaStore.MediaColumns.DATA,music.getUrl());
            //音乐路径
106         values.put(MediaStore.MediaColumns.MIME_TYPE,"audio/mp3");
107         values.put(MediaStore.Audio.Media.IS_RINGTONE,true);
            //是否铃声
108         values.put(MediaStore.Audio.Media.IS_NOTIFICATION,false);
            //是否通知声
109         values.put(MediaStore.Audio.Media.IS_ALARM,false);
            //是否闹钟声
110         values.put(MediaStore.Audio.Media.IS_MUSIC,false);      //是否音乐
111         Uri uri=MediaStore.Audio.Media.getContentUriForPath(music
                .getUrl());                              //根据路径获取对应的URI
112         Uri newUri=getActivity().getContentResolver().insert(uri,
```

```
                     values);                              //插入新的值
113        RingtoneManager.setActualDefaultRingtoneUri(getActivity(),
                RingtoneManager.TYPE_RINGTONE,newUri);
114    }
115 }
```

音乐列表 Fragment 对应的布局文件：17\MusicPlayer\res\layout\music_item.xml

```
1  <LinearLayout xmlns:android="http://schemas.android.com/apk/res/android"
2      android:layout_width="match_parent"
3      android:layout_height="match_parent"
4      android:gravity="center_vertical"
5      android:orientation="horizontal">
6      <ImageView
7          android:id="@+id/icon"
8          android:layout_width="50dp"
9          android:layout_height="50dp"
10         android:layout_margin="5dp"
11         android:contentDescription="@string/imageInfo"/>
12     <LinearLayout
13         android:layout_width="0dp"
14         android:layout_height="wrap_content"
15         android:layout_weight="1"
16         android:orientation="vertical"
17         android:padding="5dp">
18         <TextView
19             android:id="@+id/title"
20             android:singleLine="true"
21             android:ellipsize="end"
22             android:paddingRight="10dp"
23             android:textSize="18sp"
24             style="@style/textStyle"/>
25
26         <TextView
27             android:id="@+id/artist"
28             style="@style/textStyle"/>
29     </LinearLayout>
30     <TextView
31         android:id="@+id/time"
32         android:paddingRight="20dp"
33         android:textSize="18sp"
34         style="@style/textStyle"/>
35 </LinearLayout>
```

按艺术家分类 Fragment：17\MusicPlayer\src\iet\jxufe\cn\android\musicplayer\ArtistListFragment.java

```java
1    public class ArtistListFragment extends ListFragment{
2        private List<Music>musicList;      //所有音乐的集合
3        private List<MusicGroupByArtist>artistsList=new ArrayList
             <MusicGroupByArtist>();
         //所有艺术家的集合
4        public void onCreate(Bundle savedInstanceState){
5            super.onCreate(savedInstanceState);
6            musicList=MusicUtils.getMusicData(getActivity()); //获取所有音乐
7            if(musicList !=null){
8                MusicGroupByArtist(musicList);              //调用方法对音乐进行分组
9            }else{
10               Toast.makeText(getActivity(),"存储卡中暂时没有音乐,请添加音乐...",
11                   Toast.LENGTH_SHORT).show();             //提示存储卡中没有音乐
12           }
13       }
14       public void MusicGroupByArtist(List<Music>musicList){
             //对音乐进行分组,并统计每个艺术家包含的音乐数
15           for(int i=0;i <musicList.size();i++){    //循环遍历每一首音乐,判断其专辑
16               int j=0;
17               for(;j <artistsList.size();j++){
                     //循环遍历已有的艺术家,判断是否已经存在该艺术家
18                   if(musicList.get(i).getSinger()
19                       .equals(artistsList.get(j).getArtistName())){
                         //如果已经存在该艺术家,则添加一个
20                       artistsList.get(j).setCount(
21                           artistsList.get(j).getCount()+1);    //数量加1
22                       artistsList.get(j).getMusics().add(musicList.get(i));
                         //并把音乐添加到集合中
23                       break;                                    //退出循环
24                   }
25               }
26               if(j==artistsList.size()){ //如果不存在该艺术家,则将其添加进去
27                   MusicGroupByArtist artist=new MusicGroupByArtist();
                     //创建一个艺术家
28                   artist.setArtistName(musicList.get(i).getSinger());
                     //设置艺术家的名字
29                   artist.setCount(1);                      //默认歌曲数为1
30                   List<Music>musics=new ArrayList<Music>();
                     //创建一个集合用于保存该艺术家所有的音乐
31                   musics.add(musicList.get(i));            //向集合中添加音乐
32                   artist.setMusics(musics);
33                   artistsList.add(artist);                 //将艺术家添加到集合中
```

```
34              }
35          }
36      }
37      private class MusicGroupByArtist{              //艺术家分组信息
38          private String artistName;                 //艺术家的名字
39          private int count;                         //包含的音乐数量
40          private List<Music>musics;                 //具体包含的音乐集合
41          //相关属性的 set 和 get 方法
42          public String getArtistName(){
43              return artistName;
44          }
45          public void setArtistName(String artistName){
46              this.artistName=artistName;
47          }
48          public int getCount(){
49              return count;
50          }
51          public void setCount(int count){
52              this.count=count;
53          }
54          public List<Music>getMusics(){
55              return musics;
56          }
57          public void setMusics(List<Music>musics){
58              this.musics=musics;
59          }
60      }
61      public View onCreateView(LayoutInflater inflater,ViewGroup container,
62              Bundle savedInstanceState){
63          setListAdapter(new ArtistAdapter());        //显示按艺术家分类的结果
64          return super.onCreateView(inflater,container,savedInstanceState);
65      }
66      private class ArtistAdapter extends BaseAdapter{
67          public int getCount(){
68              return artistsList.size();
69          }
70          public Object getItem(int position){
71              return artistsList.get(position);
72          }
73          public long getItemId(int position){
74              return position;
75          }
76          public View getView(int position,View convertView,ViewGroup parent){
77              if(convertView==null){
```

```
78                    convertView=LinearLayout.inflate(getActivity(),
79                        R.layout.artist_item,null);
80                }
81                ImageView icon=(ImageView)convertView.findViewById(R.id.icon);
                  //显示图标的控件
82                TextView album=(TextView)convertView.findViewById(R.id.artist);
                  //显示演唱者名字
83                TextView info=(TextView)convertView.findViewById(R.id.info);
                  //显示一共有多少首歌曲
84                Bitmap bitmap=MusicUtils.getAlbumPic(getActivity(),
                        artistsList.get(position).getMusics().get(0));
85                if(bitmap!=null){
86                    icon.setImageBitmap(bitmap);
87                }else{
88                    icon.setImageResource(R.drawable.artist);
89                }
90                album.setText(artistsList.get(position).getArtistName());
91                info.setText(Html.fromHtml("共有<font color=red><b>" +
                        artistsList.get(position).getCount()+"</b></font>首歌曲"));
92                return convertView;
93            }
94        }
95        public void onListItemClick(ListView l,View v,int position,long id){
          //单击某个艺术家后显示该艺术家的所有音乐
96            MusicListFragment musicListFragment=new MusicListFragment();
97            musicListFragment.setMusicList(artistsList.get(position)
                    .getMusics());
98            FragmentTransaction fTransaction=getActivity()
                    .getFragmentManager().beginTransaction();
99            fTransaction.replace(R.id.realContent,musicListFragment);
100           fTransaction.commit();                          //提交事务
101           super.onListItemClick(l,v,position,id);
102       }
103   }
```

艺术家列表 Fragment 对应的布局文件：17\MusicPlayer\res\layout\ MusicPlayer\res\layout\artist_item.xml

```
1   <LinearLayout xmlns:android="http://schemas.android.com/apk/res/android"
2       android:layout_width="match_parent"
3       android:layout_height="match_parent"
4       android:gravity="center_vertical"
5       android:orientation="horizontal" >
6       <ImageView
7           android:id="@+id/icon"
```

```
8            android:layout_width="50dp"
9            android:layout_height="50dp"
10           android:layout_margin="5dp"
11           android:contentDescription="@string/imageInfo" />
12       <LinearLayout
13           android:layout_width="0dp"
14           android:layout_height="wrap_content"
15           android:layout_weight="1"
16           android:orientation="vertical"
17           android:padding="5dp" >
18           <TextView
19               android:id="@+id/artist"
20               android:singleLine="true"
21               android:ellipsize="end"
22               android:paddingRight="10dp"
23               android:textSize="18sp"
24               style="@style/textStyle"/>
25           <TextView
26               android:id="@+id/info"
27               style="@style/textStyle" />
28       </LinearLayout>
29   </LinearLayout>
```

按专辑分类 Fragment：17\MusicPlayer\src\iet\jxufe\cn\android\musicplayer\AlbumListFragment.java

```
1    public class AlbumListFragment extends ListFragment{
2       private List<Music>musicList;                            //所有音乐的集合
3       private List<MusicGroupByAlbum>albumList=new
           ArrayList<AlbumListFragment.MusicGroupByAlbum>();    //所有的专辑信息
4       public void onCreate(Bundle savedInstanceState){
5           super.onCreate(savedInstanceState);
6           musicList=MusicUtils.getMusicData(getActivity()); //获取所有音乐
7           if(musicList !=null){
8               MusicGroupByAlbum(musicList);            //调用方法对音乐进行分组
9           }else{
10              Toast.makeText(getActivity(),"存储卡中暂时没有音乐,请添加音乐...",
11                  Toast.LENGTH_SHORT).show();          //提示存储卡中没有音乐
12          }
13      }
14      public void MusicGroupByAlbum(List<Music>musicList){
            //对音乐按专辑分组,并统计每个专辑包含的音乐数
15          for(int i=0;i<musicList.size();i++){ //循环遍历每一首音乐,获取其专辑
16              int j=0;
17              for(;j<albumList.size();j++){ //循环遍历已有专辑,判断该专辑是否存在
```

```
18              if(musicList.get(i).getAlbum()
19                  .equals(albumList.get(j).getAlbumName())){
                    //如果已存在该专辑
20                  albumList.get(j).setCount(albumList.get(j).
                    getCount()+1);                  //数量加1
21                  albumList.get(j).getMusics().add(musicList.get(i));
                    //并把音乐添加到集合中
22                  break;                          //退出循环
23              }
24          }
25          if(j ==albumList.size()){               //如果列表中没有该专辑
26              MusicGroupByAlbum album=new MusicGroupByAlbum();
                //创建一个新专辑
27              album.setAlbumName(musicList.get(i).getAlbum());
28              album.setCount(1);                  //默认歌曲数为1
29              List<Music>musics=new ArrayList<Music>();
30              musics.add(musicList.get(i));
31              album.setMusics(musics);
32              albumList.add(album);
33          }
34      }
35  }
36  private class MusicGroupByAlbum{                //按专辑分组信息
37      private String albumName;                   //专辑名
38      private int count;                          //专辑中包含的歌曲数量
39      private List<Music>musics;                  //专辑中包含的歌曲
40      //相应的set和get方法
41      public String getAlbumName(){
42          return albumName;
43      }
44      public void setAlbumName(String albumName){
45          this.albumName=albumName;
46      }
47      public int getCount(){
48          return count;
49      }
50      public void setCount(int count){
51          this.count=count;
52      }
53      public List<Music>getMusics(){
54          return musics;
55      }
56      public void setMusics(List<Music>musicList){
57          this.musics=musicList;
```

```
58          }
59      }
60      public View onCreateView(LayoutInflater inflater,ViewGroup container,
61              Bundle savedInstanceState){
62          setListAdapter(new AlbumAdapter());          //显示所有专辑信息
63          return super.onCreateView(inflater,container,savedInstanceState);
64      }
65      private class AlbumAdapter extends BaseAdapter{
66          public int getCount(){
67              return albumList.size();
68          }
69          public Object getItem(int position){
70              return albumList.get(position);
71          }
72          public long getItemId(int position){
73              return position;
74          }
75          public View getView(int position,View convertView,ViewGroup parent){
76              if(convertView ==null){
77                  convertView=LinearLayout.inflate(getActivity(),
78                      R.layout.album_item,null);
79              }
80              ImageView icon=(ImageView)convertView.findViewById(R.id.icon);
                //显示图标的控件
81              TextView album=(TextView)convertView.findViewById(R.id.album);
                //显示专辑名称
82              TextView info=(TextView)convertView.findViewById(R.id.info);
                //显示一共有多少首歌曲
83              Bitmap bitmap=MusicUtils.getAlbumPic(getActivity(),albumList
84                  .get(position).getMusics().get(0));
85              if(bitmap !=null){
86                  icon.setImageBitmap(bitmap);
87              }else{
88                  icon.setImageResource(R.drawable.album);
89              }
90              album.setText(albumList.get(position).getAlbumName());
91              info.setText(Html.fromHtml("共有<font color=red><b>"
92                  +albumList.get(position).getCount()+"</b></font>首歌曲"));
93              return convertView;
94          }
95      }
96      public void onListItemClick(ListView l,View v,int position,long id){
            //选中某一专辑后显示该专辑中所包含的所有音乐信息
97          MusicListFragment musicListFragment=new MusicListFragment();
```

```
 98        musicListFragment.setMusicList(albumList.get(position).getMusics());
 99        FragmentTransaction fTransaction=getActivity()
               .getFragmentManager()
100            .beginTransaction();                    //开启事务
101        fTransaction.replace(R.id.realContent,musicListFragment);
102        fTransaction.commit();                      //提交事务
103        super.onListItemClick(l,v,position,id);
104    }
105 }
```

专辑列表 Fragment 对应的布局文件：17\MusicPlayer\res\layout\album_item.xml

```
 1 <LinearLayout xmlns:android="http://schemas.android.com/apk/res/android"
 2     android:layout_width="match_parent"
 3     android:layout_height="match_parent"
 4     android:gravity="center_vertical"
 5     android:orientation="horizontal" >
 6     <ImageView
 7         android:id="@+id/icon"
 8         android:layout_width="50dp"
 9         android:layout_height="50dp"
10         android:layout_margin="5dp"
11         android:contentDescription="@string/imageInfo" />
12     <LinearLayout
13         android:layout_width="0dp"
14         android:layout_height="wrap_content"
15         android:layout_weight="1"
16         android:orientation="vertical"
17         android:padding="5dp" >
18         <TextView
19             android:id="@+id/album"
20             android:singleLine="true"
21             android:ellipsize="end"
22             android:paddingRight="10dp"
23             android:textSize="18sp"
24             style="@style/textStyle"/>
25         <TextView
26             android:id="@+id/info"
27             style="@style/textStyle" />
28     </LinearLayout>
29 </LinearLayout>
```

播放列表 Fragment：17\MusicPlayer\src\iet\jxufe\cn\android\musicplayer\PlayListFragment.java

```
 1 public class PlayListFragment extends ListFragment{
```

```
2     public List<Music>musicList=Constants.playlist;          //播放列表中的音乐
3     private PlayListAdapter adapter;
4     public View onCreateView(LayoutInflater inflater,ViewGroup container,
5           Bundle savedInstanceState){
6        if(musicList !=null&&musicList.size()!=0){
7           adapter=new PlayListAdapter();
8           setListAdapter(adapter);
9        }else{
10         Toast.makeText(getActivity(),"播放列表中暂时没有音乐,请添加音乐...",
11             Toast.LENGTH_SHORT).show();
12       }
13       return super.onCreateView(inflater,container,savedInstanceState);
14    }
15    public void onStart(){
16       registerForContextMenu(getListView());   //为音乐列表注册上下文菜单
17       super.onStart();
18    }
19    private class PlayListAdapter extends BaseAdapter{
20       public int getCount(){
21          return musicList.size();
22       }
23       public Object getItem(int position){
24          return musicList.get(position);
25       }
26       public long getItemId(int position){
27          return position;
28       }
29       public View getView(int position,View convertView,ViewGroup parent){
30          if(convertView ==null){
31             convertView =LinearLayout.inflate(getActivity(),
32                R.layout.music_item,null);
33          }
34          ImageView icon = (ImageView)convertView.findViewById(R.id.icon);
             //显示图标的控件
35          TextView title = (TextView)convertView.findViewById(R.id.title);
             //显示歌曲名的控件
36          TextView artist = (TextView)convertView.findViewById
                (R.id.artist);                              //显示演唱者的控件
37          TextView time = (TextView)convertView.findViewById(R.id.time);
             //显示时间的控件
38          Bitmap bitmap=MusicUtils.getAlbumPic(getActivity(),
                musicList.get(position));
39          if(bitmap!=null){
40             icon.setImageBitmap(bitmap);
```

```
41              }else{
42                  icon.setImageResource(R.drawable.music);
43              }
44              title.setText(musicList.get(position).getTitle());
45              artist.setText(musicList.get(position).getSinger());
46              time.setText(MusicUtils.timeToString(musicList.get(position)
47                  .getTime()));
48              return convertView;
49          }
50      }
51      public void onListItemClick(ListView l,View v,int position,long id){
            //音乐项被单击事件处理
52          Intent intent =new Intent(getActivity(),MusicPlayActivity.class);
53          intent.putExtra("listType",Constants.PLAY_LIST_MUSIC);
            //音乐播放的列表类型:播放列表
54          intent.putExtra("music",musicList.get(position));     //当前播放的音乐
55          intent.putExtra("position",position);        //当前音乐在列表中的索引
56          startActivity(intent);
57          super.onListItemClick(l,v,position,id);
58      }
59      public void onCreateContextMenu(ContextMenu menu,View v,
60              ContextMenuInfo menuInfo){              //创建上下文菜单
61          getActivity().getMenuInflater().inflate(R.menu.playlist_
                context,menu);
62          super.onCreateContextMenu(menu,v,menuInfo);
63      }
64      public boolean onContextItemSelected(MenuItem item){
65          AdapterContextMenuInfo info=(AdapterContextMenuInfo)item
                .getMenuInfo();
66          switch(item.getItemId()){
67          case R.id.deleteAll:                        //清空播放列表
68              musicList.clear();
69              System.out.println(musicList);
70              break;
71          case R.id.deleteFromList:                   //从播放列表中删除
72              musicList.remove(info.position);
73              break;
74          default:
75              break;
76          }
77          adapter.notifyDataSetChanged();
78          Constants.playlist=musicList;
79          return super.onContextItemSelected(item);
80      }
```

```
81  }
```

常量类：17\MusicPlayer\src\iet\jxufe\cn\android\musicplayer\Constants.java

```
1   public class Constants{
2       public static List<Music>musiclist=new ArrayList<Music>();
        //所有音乐的集合
3       public static List<Music>playlist=new ArrayList<Music>();//音乐播放列表
4       public static final String CONTROL_ACTION="iet.jxufe.cn.android.control";
        //控制音乐播放动作,即播放或暂停
5       public static final String SEEKBAR_ACTION="iet.jxufe.cn.android.seekbar";
        //音乐进度发送变化动作
6       public static final String COMPLETE_ACTION="iet.jxufe.cn.android.complete";
        //音乐播放结束动作
7       public static final String UPDATE_ACTION="iet.jxufe.cn.android.update";
        //更新进度条
8       public static final String UPDATE_STYLE="iet.jxufe.cn.android.style";
        //更新播放形式
9       public static final Uri ALBUM_URL=Uri.parse
                    ("content://media/external/audio/albumart");
10      public static final String LIST_LOOP="列表循环";
11      public static final String SINGLE_LOOP="单曲循环";
12      public static final String OVER_FINISH="结束后停止";
13      public static final String RANDOM_PLAY="随机播放";
14      public static final int NEW=6;              //开始一首新的音乐
15      public static final int PLAY=1;             //播放
16      public static final int PAUSE=2;            //暂停
17      public static final int ALL_MUSIC=0x11;     //播放所有的音乐
18      public static final int PLAY_LIST_MUSIC=0x12;   //播放播放列表中的音乐
19  }
```

控制音乐播放的Activity：17\MusicPlayer\src\iet\jxufe\cn\android\musicplayer\MusicPlayActivity.java

```
1   public class MusicPlayActivity extends Activity{
        //音乐播放主界面,可控制音乐的播放
2       private List<Music>musicList;               //记录音乐列表
3       private TextView titleView,singerView,currentTimeView,totalTimeView;
        //显示歌曲名、作者信息、当前播放时间、总时间的文本框
4       private SeekBar playProgress;               //拖动条
5       private Spinner styleSpinner;               //选择播放形式的下拉列表
6       private ImageButton control;                //播放/暂停按钮
7       private ImageView picView;                  //显示图片
8       private ServerReceiver serverReceiver;      //接收后台服务发送的广播的广播接收器
9       private Music currentMusic;                 //记录当前播放的音乐
```

```java
10    private boolean isPause=false;          //是否暂停
11    private int currentPosition;            //当前音乐的索引
12    private int listType;    //音乐列表的类型,是播放所有音乐还是播放列表中的音乐
13    private String[] styles=new String[]{Constants.LIST_LOOP,Constants
          .SINGLE_LOOP,Constants.RANDOM_PLAY,Constants.OVER_FINISH};
14    protected void onCreate(Bundle savedInstanceState){
15        super.onCreate(savedInstanceState);
16        requestWindowFeature(Window.FEATURE_NO_TITLE);    //去除标题
17        getWindow().setFlags(WindowManager.LayoutParams.FLAG_FULLSCREEN,
18            WindowManager.LayoutParams.FLAG_FULLSCREEN);    //全屏显示
19        setContentView(R.layout.play_item);               //加载主界面
20        initView();                                        //初始化界面
21    }
22    public void initView(){                                //执行初始化操作
23        styleSpinner= (Spinner)findViewById(R.id.styleSpinner);
24        styleSpinner.setAdapter(new ArrayAdapter<String>(this,
              R.layout.spinner_text,styles));
25        styleSpinner.setOnItemSelectedListener(new
              SpinnerItemClickListener());
26        picView= (ImageView)findViewById(R.id.picView);
27        currentTimeView =(TextView)findViewById(R.id.currentTime);
      //显示音乐当前播放的时间的文本控件
28        totalTimeView =(TextView)findViewById(R.id.totalTime);
      //显示音乐总时长的文本控件
29        titleView =(TextView)findViewById(R.id.title);
      //显示音乐标题的文本控件
30        singerView =(TextView)findViewById(R.id.singer);
      //显示音乐演唱者的文本控件
31        playProgress =(SeekBar)findViewById(R.id.playProgress);
      //显示当前音乐播放进度的控件
32        control = (ImageButton)findViewById(R.id.control);
      //控制播放或暂停的控件
33        playProgress.setOnSeekBarChangeListener(new
              MySeekBarChangeListener());          //为拖动条添加事件处理
34        serverReceiver =new ServerReceiver();    //创建广播接收器
35        IntentFilter filter =new IntentFilter(); //可以接收到的广播类型
36        filter.addAction(Constants.COMPLETE_ACTION); //音乐播放结束的事件
37        filter.addAction(Constants.UPDATE_ACTION);   //更新进度的动作
38        registerReceiver(serverReceiver,filter);     //注册广播接收器
39        currentMusic = (Music)getIntent().getSerializableExtra
              ("music");                              //获取当前播放的音乐
40        if(currentMusic==null){                     //如果当前音乐为空
41            SharedPreferences musicPreferences=getSharedPreferences
                  ("music",Context.MODE_PRIVATE);
```

```
42          currentPosition=musicPreferences.getInt("position",0);
43          listType=musicPreferences.getInt("listType",
                Constants.ALL_MUSIC);
44          if(listType==Constants.ALL_MUSIC){      //如果是所有的音乐
45              musicList=Constants.musiclist;
46          }else{                                   //如果是播放列表
47              musicList=Constants.playlist;
48          }
49          currentMusic=musicList.get(currentPosition);    //获取当前的音乐
50          String styleString=musicPreferences.getString("style",
                Constants.LIST_LOOP);
51          for(int i=0;i<styles.length;i++){
52              if(styles[i].equalsIgnoreCase(styleString)){
53                  styleSpinner.setSelection(i);
54                  break;
55              }
56          }
57          showInfo();                              //显示音乐信息
58          isPause=true;                            //继续播放
59          control(null);
60      }else{                                        //直接播放音乐
61          currentPosition=getIntent().getIntExtra("position",0);
            //默认为第一首
62          listType=getIntent().getIntExtra("listType",
                Constants.ALL_MUSIC);
            //获取列表的类型,是播放所有音乐还是从播放列表中播放
63          playNewMusic();                          //播放音乐
64          if(listType==Constants.ALL_MUSIC){      //如果是播放所有音乐
65              musicList=Constants.musiclist;
66          }else{                                   //如果是播放列表
67              musicList=Constants.playlist;
68          }
69      }
70  }
71  public void showInfo(){
72      totalTimeView.setText(MusicUtils.timeToString
            (currentMusic.getTime()));                //显示音乐的总时长
73      titleView.setText("    "+currentMusic.getTitle()+"    ");
        //显示歌曲名
74      singerView.setText(currentMusic.getSinger());   //显示演唱者
75      Bitmap bitmap=MusicUtils.getAlbumPic(this,currentMusic);
76      if(bitmap!=null){
77          picView.setImageBitmap(bitmap);
78      }else{
```

```java
79              picView.setImageResource(R.drawable.background);
80          }
81      }
82      public void playNewMusic(){
83          currentTimeView.setText(MusicUtils.timeToString(0));
            //显示当前播放的时间,默认为 0
84          showInfo();
85          playProgress.setProgress(0);                        //进度为 0
86          control.setImageResource(R.drawable.pause);         //显示暂停的按钮
            //发送广播,通知后台播放音乐
87
88          Intent controlIntent =new Intent(Constants.CONTROL_ACTION);
89          controlIntent.putExtra("new",Constants.NEW); //这是一首新音乐
90          controlIntent.putExtra("position",currentPosition);
            //传递当前音乐的序号
91          controlIntent.putExtra("listType",listType);    //传递音乐列表的类型
92          sendBroadcast(controlIntent);                   //发送广播
93          isPause=false;                                  //是否暂停为 false
94      }
95      private class MySeekBarChangeListener implements OnSeekBarChangeListener{
96          public void onProgressChanged(SeekBar seekBar,int progress,
97              boolean fromUser){              //进度发送变化时调用该方法
98          }
99          public void onStartTrackingTouch (SeekBar seekBar){  //开始拖动时调用
100         }
101         public void onStopTrackingTouch (SeekBar seekBar){   //结束拖动时调用
102             Intent seekIntent =new Intent(Constants.SEEKBAR_ACTION);
                //发送广播通知拖动条变化
103             seekIntent.putExtra("progress",seekBar.getProgress());
                //将当前的进度传递进去
104             sendBroadcast(seekIntent);                      //发送广播
105         }
106     }
107     private class ServerReceiver extends BroadcastReceiver{
            //广播接收器,用于接收后台服务发送的广播
108         public void onReceive(Context context,Intent intent){
109             if(intent.getAction()==Constants.UPDATE_ACTION){
                    //更新进度的广播处理
110                 int position=intent.getIntExtra("position",0);
                    //获取音乐播放的位置
111                 currentTimeView.setText(MusicUtils.timeToString
                        (position));                    //显示当前的播放时长
112                 playProgress.setProgress((int)(position*1.0/
                        currentMusic.getTime()*100));   //根据位置计算进度条的进度
113             }else if(intent.getAction()==Constants.COMPLETE_ACTION){
```

```java
                    //音乐播放完成的事件处理
114                 currentPosition=intent.getIntExtra("position",0);
                    //获取当前播放的音乐的序号
115                 currentMusic=musicList.get(currentPosition);
                    //获取当前的音乐
116                 showInfo();                         //显示当前音乐的信息
117             }
118         }
119     }
120     private class SpinnerItemClickListener implements
            OnItemSelectedListener{
121         public void onItemSelected(AdapterView<?>parent,View view,
122                 int position,long id){
123             Intent styleIntent=new Intent(Constants.UPDATE_STYLE);
124             styleIntent.putExtra("style",styles[position]);
125             sendBroadcast(styleIntent);
126         }
127         public void onNothingSelected(AdapterView<?>parent){
128         }
129     }
130     public void chooseMusic(View view){              //选择歌曲按钮的事件处理
131         Intent intent=new Intent(this,MainActivity.class);
132         startActivity(intent);
133         this.finish();
134     }
135     public void first(View view){                    //第一首按钮的事件处理
136         currentPosition=0;
137         currentMusic=musicList.get(currentPosition);
138         playNewMusic();
139     }
140     public void pre(View view){                      //前一首按钮的事件处理
141         currentPosition=(currentPosition-1+musicList.size())%
                musicList.size();
142         currentMusic=musicList.get(currentPosition);
143         playNewMusic();
144     }
145     public void control(View view){                  //播放和暂停按钮的事件处理
146         Intent controlIntent=new Intent(Constants.CONTROL_ACTION);
            //控制音乐的播放和暂停
147         if(!isPause){                                //如果处于播放状态,发送广播通知暂停
148             controlIntent.putExtra("control",Constants.PAUSE);
149             control.setImageResource(R.drawable.play);   //改变图标
150         }else{                                       //如果处于暂停状态,发送广播通知播放
151             controlIntent.putExtra("control",Constants.PLAY);
```

```
152             control.setImageResource(R.drawable.pause);     //改变图标
153         }
154         isPause=!isPause;
155         sendBroadcast(controlIntent);                         //发送广播
156     }
157     public void next(View view){          //下一首按钮的事件处理
158         currentPosition=(currentPosition+1)%musicList.size();
159         currentMusic=musicList.get(currentPosition);
160         playNewMusic();
161     }
162     public void last(View view){          //最后一首按钮的事件处理
163         currentPosition=musicList.size()-1;
164         currentMusic=musicList.get(currentPosition);
165         playNewMusic();
166     }
167     protected void onDestroy(){           //服务销毁时取消广播接收器的注册
168         if(serverReceiver !=null){
169             unregisterReceiver(serverReceiver); //Activity销毁时取消注册
170         }
171         super.onDestroy();
172     }
173 }
```

播放音乐的后台Service:17\MusicPlayer\src\iet\jxufe\cn\android\musicplayer\MusicService.java

```
1   public class MusicService extends Service{
2       private List<Music>musicList;              //音乐列表
3       private int position;                      //当前音乐的序号
4       private MediaPlayer mediaPlayer;           //媒体播放器
5       private ActivityReceiver activityReceiver; //广播接收器
6       private Music currentMusic;                //当前播放的音乐
7       private Timer timer;                       //定时器
8       private int listType;                      //列表的类型
9       private String styleString=Constants.LIST_LOOP; //音乐播放形式,默认列表循环
10      public void onCreate(){                    //启动服务
11          mediaPlayer=new MediaPlayer();
12          activityReceiver=new ActivityReceiver();
13          IntentFilter filter=new IntentFilter(); //创建Intent过滤器
14          filter.addAction(Constants.CONTROL_ACTION);
            //控制音乐播放的动作,包括播放和暂停
15          filter.addAction(Constants.SEEKBAR_ACTION);  //改变音乐播放进度
16          filter.addAction(Constants.UPDATE_STYLE);    //改变音乐播放形式
17          registerReceiver(activityReceiver,filter);   //注册广播接收器
18          super.onCreate();
```

```java
19      }
20      public IBinder onBind(Intent intent){
21          return null;
22      }
23      private class ActivityReceiver extends BroadcastReceiver{
                //获取前台发送的广播
24          public void onReceive(Context context,Intent intent){
25              if(intent.getAction()==Constants.CONTROL_ACTION){
                    //接收到控制播放的广播(开始新音乐、暂停、播放)
26                  int isNew=intent.getIntExtra("new",-1);
27                  if(isNew!=-1){                              //播放一首新音乐
28                      listType=intent.getIntExtra("listType",
                            Constants.ALL_MUSIC);
29                      if(listType==Constants.ALL_MUSIC){   //如果是所有的音乐
30                          musicList=Constants.musiclist;
31                      }else{                                  //如果是播放列表
32                          musicList=Constants.playlist;
33                      }
34                      position=intent.getIntExtra("position",0);
35                      currentMusic=musicList.get(position);
                        //获取当前需要播放的音乐
36                      preparedAndPlay(currentMusic);         //准备并播放音乐
37                  }else{
38                      int control=intent.getIntExtra("control",-1);
39                      if(control==Constants.PAUSE){          //表示要暂停音乐
40                          mediaPlayer.pause();               //音乐暂停
41                          timer.cancel();                    //取消定时器
42                      }else if(control==Constants.PLAY){     //表示继续播放音乐
43                          mediaPlayer.start();
44                          startTimer();                      //启动定时器
45                      }
46                  }
47              }else if(intent.getAction()==Constants.SEEKBAR_ACTION){
48                  int progress=intent.getIntExtra("progress",0);   //获取传递的进度
49                  int position=(int)(currentMusic.getTime() * progress *
                        1.0/100);                          //将进度转换成相应的时间位置
50                  mediaPlayer.seekTo(position);    //音乐跳转到指定位置继续播放
51              }else if(intent.getAction()==Constants.UPDATE_STYLE){
52                  styleString=intent.getStringExtra("style");
53                  SharedPreferences musicPreferences=getSharedPreferences
                        ("music",Context.MODE_PRIVATE);
54                  Editor editor=musicPreferences.edit();     //获取参数编辑器
55                  editor.putString("style",styleString);
56                  editor.commit();                           //提交数据
```

```
57              }
58          }
59      }
60      public void preparedAndPlay(Music music){           //准备并播放音乐
61          try{
62              mediaPlayer.reset();                         //重置媒体播放器
63              mediaPlayer.setDataSource(music.getUrl());   //设置音乐播放的路径
64              mediaPlayer.prepare();                       //准备播放音乐
65              mediaPlayer.start();                         //播放音乐
66              startTimer();                                //启动定时器
67              sendNotification();                          //发送广播
68              saveInfo();                                  //保存数据
69              mediaPlayer.setOnCompletionListener(new OnCompletionListener(){
                //音乐播放结束事件监听器
70                  public void onCompletion(MediaPlayer mp){
                    //音乐播放完成后根据设置的播放类型进行播放,并通知前台改变
71                      if(!Constants.OVER_FINISH.equalsIgnoreCase
                        (styleString)){
                            //如果不是播放结束停止
72                          if(Constants.LIST_LOOP.equalsIgnoreCase
                            (styleString)){
73                              position=(position+1)%musicList.size();
                                //自动播放下一首
74                          }else if(Constants.RANDOM_PLAY.equalsIgnoreCase
                            (styleString)){
75                              position=new Random().nextInt(musicList.size());
76                          }
77                          currentMusic=musicList.get(position); //获取当前的音乐
78                          preparedAndPlay(currentMusic); //准备并播放音乐
79                          Intent intent=new Intent(Constants.COMPLETE_ACTION);
80                          intent.putExtra("position",position);
81                          sendBroadcast(intent);               //发送广播
82                      }else{
83                          stopSelf();                          //结束
84                      }
85                  }
86              });
87          }catch(Exception ex){
88              ex.printStackTrace();
89          }
90      }
91      public void saveInfo(){                              //保存信息
92          SharedPreferences musicPreferences=getSharedPreferences("music",
            Context.MODE_PRIVATE);
```

```
93          Editor editor=musicPreferences.edit();            //获取参数编辑器
94          editor.putInt("listType",listType);               //保存音乐列表类型
95          editor.putInt("position",position);               //保存音乐的位置
96          editor.commit();                                  //提交数据
97      }
98      public void sendNotification(){                       //后台发送通知
99          NotificationManager notificationManager=(NotificationManager)
                getSystemService(Service.NOTIFICATION_SERVICE); //获取通知服务器
100         Builder builder=new Notification.Builder(this);   //通知构建器
101         builder.setAutoCancel(false);        //打开通知后自动消除为 false
102         builder.setTicker("音乐播放");        //显示在状态栏上的通知提示信息
103         builder.setSmallIcon(R.drawable.music);           //设置通知的小图标
104             builder. setLargeIcon ( BitmapFactory. decodeResource
                (getResources(),R.drawable.largeicon));
                //设置通知的大图标
105         builder.setContentTitle("正在播放音乐");          //设置通知内容的标题
106         builder.setContentText(currentMusic.getTitle()+" "+
                currentMusic.getSinger());                    //设置通知的内容
107         Intent intent=new Intent("iet.jxufe.cn.android.music_play");
            //通知启动的页面
108         PendingIntent pIntent=PendingIntent.getActivity(this,0,intent,0);
109         builder.setContentIntent(pIntent);   //设置通知启动的程序
110         notificationManager.notify(0x11,builder.build());       //发送通知
111     }
112     public void startTimer(){              //启动定时器
113         timer=new Timer();                 //创建定时器对象
114         timer.schedule(new TimerTask(){    //定时执行的任务
115             public void run(){             //发送广播,通知更新前台进度条
116                 Intent updateIntent=new Intent(Constants.UPDATE_ACTION);
117                 updateIntent.putExtra("position",mediaPlayer
                        .getCurrentPosition());
118                 sendBroadcast(updateIntent);
119             }
120         },0,1000);                         //每隔 1s 发一次
121     }
122     public void onDestroy(){               //服务销毁时调用该方法
123         if(mediaPlayer!=null){             //重置音乐播放器
124             mediaPlayer.reset();
125         }
126         if(activityReceiver!=null){        //取消广播接收器
127             unregisterReceiver(activityReceiver);
128         }
129         super.onDestroy();
130     }
```

131   }

样式文件：17\MusicPlayer\res\values\styles.xml

```xml
1   <resources>
2       <style name="AppBaseTheme" parent="android:Theme.Light">
3       </style>
4       <style name="AppTheme" parent="AppBaseTheme">
5       </style>
6       <style name="textStyle">
7           <item name="android:layout_width">wrap_content</item>
8           <item name="android:layout_height">wrap_content</item>
9           <item name="android:textColor">#ffffff</item>
10          <item name="android:textSize">14sp</item>
11      </style>
12      <style name="imageBtnStyle">
13          <item name="android:layout_width">wrap_content</item>
14          <item name="android:layout_height">wrap_content</item>
15          <item name="android:background">#00000000</item>
16          <item name="android:layout_marginRight">10dp</item>
17      </style>
18  </resources>
```

清单文件：17\MusicPlayer\AndroidManifest.xml

```xml
1   <?xml version="1.0" encoding="utf-8"?>
2   <manifest xmlns:android="http://schemas.android.com/apk/res/android"
3       package="iet.jxufe.cn.android.musicplayer"
4       android:versionCode="1"
5       android:versionName="1.0">
6       <uses-sdk
7           android:minSdkVersion="16"
8           android:targetSdkVersion="17"/>
9       <application
10          android:allowBackup="true"
11          android:icon="@drawable/ic_launcher"
12          android:label="@string/app_name"
13          android:theme="@style/AppTheme">
14          <activity
15              android:name="iet.jxufe.cn.android.musicplayer.MainActivity"
16              android:label="@string/app_name"
17              android:launchMode="singleInstance">
18              <intent-filter>
19                  <action android:name="android.intent.action.MAIN"/>
```

```
20              <category android:name="android.intent.category.LAUNCHER"/>
21          </intent-filter>
22      </activity>
23      <activity android:name="iet.jxufe.cn.android.musicplayer
            .MusicPlayActivity">
24          <intent-filter>
25              <action android:name="iet.jxufe.cn.android.music_play"/>
26              <category android:name="android.intent.category.DEFAULT"/>
27          </intent-filter>
28      </activity>
29      <service android:name="iet.jxufe.cn.android.musicplayer
            .MusicService"></service>
30  </application>
31  <uses-permission android:name="android.permission.MOUNT_UNMOUNT_
        FILESYSTEMS"/>
32  <uses-permission android:name="android.permission.READ_EXTERNAL_
        STORAGE"/>
33  <uses-permission android:name="android.permission.WRITE_EXTERNAL_
        STORAGE"/>
34  <!--真机测试需要加上该权限,否则会报错,即"refusing to reopen boot dex '/
        system/framework/hwframework.jar'"-->
35  <uses-permission android:name="android.permission.WRITE_SETTINGS"/>
36  </manifest>
```

17.3 代码分析

17.3.1 音乐播放器的主要功能分析

本案例实现了音乐播放器的一些常用功能,主要包括以下功能:

(1) 获取存储卡中所有 MP3 格式的音乐,并以列表的形式显示。

(2) 将音乐进行分类,例如按艺术家和按专辑分类,并进行简单统计,统计每类中包含的音乐数量,单击某一项后显示该类别下所有的音乐。

(3) 实现播放列表功能,用户可以将自己喜欢的歌曲添加到播放列表中,在播放时可以从播放列表中开始播放音乐。

(4) 将某一首音乐设置为手机铃声。

(5) 显示当前音乐的播放时间和进度,支持拖动滚动条改变音乐的播放进度。

(6) 能够控制音乐的播放、暂停,实现第一首、上一首、下一首、最后一首功能。

(7) 支持多种音乐播放形式,例如列表循环、单曲循环、随机播放、结束后停止等。

音乐播放器案例的主要页面跳转和功能流程如图 17-10 所示。

按艺术家分类、音乐列表、按专辑分类、播放列表之间的切换主要是通过 TabHost+Fragment 来实现的,整体只有一个 Activity——MainActivity。获取存储卡中所有的音乐信息主要是调用系统提供的 ContentProvider,然后对音乐信息进行分类统计。

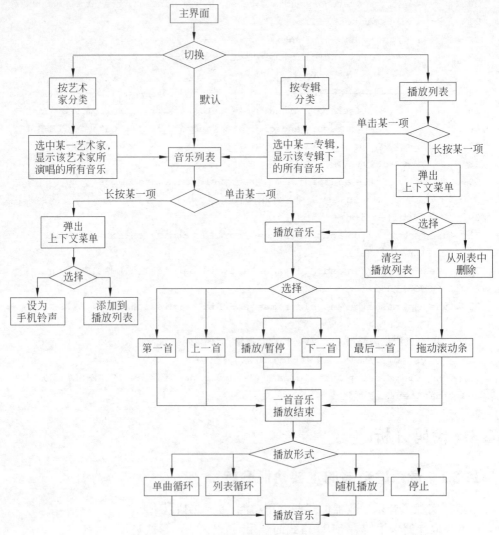

图 17-10　音乐播放器的主要功能和流程

为音乐列表添加单击事件处理和上下文菜单，单击某一项后，跳转到音乐播放界面；长按某一项后，弹出上下文菜单，可将该音乐设置为手机铃声，也可以将该音乐添加到播放列表中。在添加到播放列表时先判断播放列表中是否存在该音乐，如果已经存在则不添加。

播放列表功能，为了保存用户的播放列表，不用每次都重新添加，应该在退出程序时将播放列表中的信息保存到本地文件中；在启动时，再从本地文件中获取相关的信息。由于音乐信息是一种结构化的数据，本案例采用 SQLite 数据库来保存列表中的音乐信息。

17.3.2　Android 四大组件之 ContentProvider

ContentProvider 是不同应用程序之间进行数据交换的标准 API，为存储和读取数据提供了统一的接口。通过 ContentProvider，应用程序可以实现数据共享。Android 内置的许多应用都使用 ContentProvider 向外提供数据，供开发者调用（如视频、音频、图片、通讯录等），其中最典型的应用就是通讯录。

那么，ContentProvider 是如何对外提供数据的呢，又是如何实现这一机制的呢？ContentProvider 以某种 URI 的形式对外提供数据，数据以类似数据库中表的方式暴露，允许其他应用访问或修改数据，其他应用程序使用 ContentResolver 根据 URI 去访问操作指定的数据。URI 是通用资源标识符，即每个 ContentProvider 都有一个唯一标识的 URI，其他应用程序的 ContentResolver 根据 URI 就知道具体解析的是哪个 ContentProvider，然后调用相应的操作方法，而 ContentResolver 的方法内部实际上是调用该 ContentProvider 的对应方法，对于 ContentProvider 方法内部是如何实现的，其他应用程序是不知道具体细节的，只是知道有那个方法，这就达到了统一接口的目的。对于不同的数据存储方式，该方法内部的实现是不同的，而外部访问方法都是一致的。

ContentProvider 也是 Android 四大组件之一，如果用户要开发自己的 ContentProvider，必须实现 Android 系统提供的 ContentProvider 基类，并且需要在 AndroidManifest.xml 文件中进行配置。ContentProvider 基类的常用方法如下。

- public **abstract boolean onCreate**()：该方法在 ContentProvider 创建后调用，当其他应用程序第一次访问 ContentProvider 时，ContentProvider 会被创建，并立即调用该方法。
- public **abstract Cursor query**(Uri uri，String[] projection，String selection，String[] selectionArgs，String sortOrder)：根据 Uri 查询符合条件的全部记录，其中，projection 是所需要获取的数据列，selection 表示查询条件，selectionArgs 表示查询条件中的占位符对应的值，sortOrder 表示查询结果的排序。
- public **abstract int update**(Uri uri，ContentValues values，String select，String[] selectArgs)：根据 Uri 修改 select 条件所匹配的全部记录。
- public **abstract int delete**(Uri uri，String selection，String[] selectionArgs)：根据 Uri 删除符合条件的全部记录。
- public **abstract Uri insert**(Uri uri，ContentValues values)：根据 Uri 插入 values 对应的数据，ContentValues 类似于 map，存放的是键值对。
- public **abstract String getType**(Uri uri)：该方法返回当前 Uri 所代表的数据的 MIME 类型。如果该 Uri 对应的数据包含多条记录，则 MIME 类型字符串应该以 vnd.android.curor.dir/开头，如果该 Uri 对应的数据只包含一条记录，则 MIME 类型字符串应该以 vnd.android.cursor.item/开头。

以上几个方法都是抽象方法，用户在开发自己的 ContentProvider 时必须重写这些方法，然后在 AndroidManifest.xml 文件中配置该 ContentProvider，为了能让其他应用找到该 ContentProvider，ContentProvider 采用了 authorities（主机名/域名）对它进行唯一标识，可以把 ContentProvider 看作是一个网站，authorities 就是它的域名，只需在 <application.../>元素内添加以下代码即可。

```
1  <provider android:name=".MyProvider"
2      android:authorities="iet.jxufe.cn.android.provider.myprovider">
```

```
3    </provider>
```

注意：authorities 是必备属性，如果没有 authorities 属性会报错。

一旦某个应用程序通过 ContentProvider 暴露了自己的数据操作接口，那么不管该应用程序是否启动，其他应用程序都可以通过该接口操作该应用程序的内部数据。

调用其他应用程序通过 ContentProvider 暴露的数据主要有下面 3 个步骤：

（1）获取该应用程序所暴露的数据对应的 URI。

（2）调用当前 Context 对象的 getContentResolver() 方法获取 ContentResolver 对象。

（3）调用 ContentResolver 对象的增、删、查、改方法对暴露的数据进行操作。

本案例主要是调用系统中为音/视频提供的 ContentProvider 来获取设备中的音/视频信息。系统提供的获取手机外部存储卡（SDCard）中的音乐信息对应的 URI 为 "MediaStore.Audio.Media.EXTERNAL_CONTENT_URI"，获取音乐信息的代码见 MusicUtils.java 的第 2~46 行。获取专辑对应的图片的 URI 为 "content://media/external/audio/albumart"，具体操作见 MusicUtils.java 的第 71~90 行。设置铃声也是通过 ContentProvider 完成的，具体代码见 MusicListFragment.java 的第 103~114 行。

17.3.3　Android 四大组件之 Service

Service（服务）与 Activity 类似，都是 Android 的四大组件之一，并且二者都是从 Context 类派生而来，最大的区别在于 Service 没有实际的界面，一直在后台运行，相当于一个没有图形界面的 Activity 程序。Service 主要有两种用途，即后台运行和跨进程访问。通过启动一个服务，可以在不显示界面的前提下在后台运行指定的任务，这样可以不影响用户做其他事情，例如后台播放音乐，前台显示网页信息。通过 AIDL 服务可以方便地实现不同进程之间的通信。

Service 本身不能直接运行，需要借助 Context 对象。启动 Service 主要有两种方式，一是调用 Context 对象的 startService() 方法启动，另一种是调用 Context 对象的 bindService() 方法绑定。在通过第一种方式启动时，启动者与 Service 之间没有关联，该 Service 将一直在后台执行，即使调用 startService() 的对象销毁了，Service 仍然存在，直到有进程调用 stopService()，或者 Service "自杀"（stopSelf()）。在这种情况下，Service 与访问者之间无法进行通信、数据交换，往往用于执行单一操作，并且没有返回结果。例如，通过网络上传、下载文件，操作一旦完成，服务应该自动销毁。通过第二种方式绑定后，Service 就和调用 bindService() 的 Context 对象同生共死了，若调用 bindService() 的对象销毁了，那么它绑定的 Service 也要跟着被结束，当然期间也可以调用 unbindservice() 让 Service 提前解绑。

注意：一个服务可以与多个对象绑定，只有当所有的对象都与之解绑后，该服务才会被销毁。

以上两种方式也可以混合使用，即一个 Service 既可以被某个 Context 对象启动，也可以与其他 Context 对象绑定，此时只有调用 stopService()，并且调用 unbindservice() 方法后，该 Service 才会被销毁。

注意：虽然服务用于执行一些耗时的操作，但服务仍运行在它所在进程的主线程，并没有创建自己的线程，也没有运行在一个独立的进程上，这意味着，如果你的服务需要做一些消耗 CPU 或者阻塞的操作，你应该在服务中创建一个新的线程去处理。

通过使用独立的线程，可以降低程序出现 ANR(Application No Response，程序没有响应)的风险，程序的主线程仍然可以与用户进行流畅的交互。

与创建 Activity 类似，在开发 Service 时需要继承 Android 系统为我们提供的 Service 基类，然后根据需要实现一些回调方法。系统中 Service 类的主要方法如下。

- abstract IBinder onBind(Intent intent)：该方法是一个抽象方法，因此所有 Service 的子类必须实现该方法。该方法将返回一个 IBinder 对象，应用程序可通过该对象与 Service 组件通信。
- void onCreate()：当 Service 第一次被创建时将立即回调该方法。
- void onDestroy()：在 Service 被关闭之前将回调该方法。
- void onStartCommand(Intent intent, int flags, int startId)：该方法的早期版本是 void onStart(Intent intent, int startId)，每次客户端调用 startService(Intent intent)方法启动该 Service 时都会回调该方法。
- boolean onUnbind(Intent intent)：当该 Service 上绑定的所有客户端都断开连接时将会回调该方法。

自定义的 Service 子类必须实现 onBind()方法，本案例中的 MusicService 类还实现了 onCreate()方法(在该方法中执行一些初始化的操作)和 onDestroy()方法(在该方法中执行一些扫尾工作)。然后还需要在 AndroidManifest.xml 文件中对该 Service 子类进行配置，在配置时可通过＜intent-filter.../＞元素指定它可被哪些 Intent 启动。例如本案例中 Service 是被显式启动的，而不是按条件启动，所以只需简单地指定 Service 的完整类名。

```
<service android:name="iet.jxufe.cn.android.musicplayer.MusicService">
</service>
```

在本案例中，当退出音乐应用后，希望仍然能够播放音乐，因此通过 startService()方法来启动 Service。音乐启动后，Service 与启动它的 Context 之间没有任何关系。这样就带来一个问题，当我们需要通过前台来控制音乐的播放时，后台如何知道呢？也就是说，此时我们如何实现前后台间的交互呢？为了解决这个问题，Android 为我们提供了一种机制——广播。

17.3.4 Android 四大组件之 BroadcastReceiver

广播是一种广泛运用在应用程序之间传输信息的机制，而 BroadcastReceiver 是对发送出来的广播进行过滤接收并响应的一类组件。BroadcastReceiver 本质上是一种全局监听器，用于监听系统全局的广播消息，因此它可以非常方便地实现系统中不同组件之间的通信。

BroadcastReceiver 用于接收指定 Intent 的广播，而广播的发送是通过 Context 对象的 sendBroadcast()以及 sendOrderedBroadcast()来实现的。通常，一个广播 Intent 可以被订阅了该 Intent 的多个广播接收者所接收，如同一个广播台可以被多位听众收听一样。

　　BroadcastReceiver 自身并不实现图形用户界面，但是当它收到某个消息后可以启动 Activity 作为响应，或者通过 NotificationManager 提醒用户，或者启动 Service 等。启动 BroadcastReceiver 与启动 Activity、Service 非常相似，需要以下两步。

　　（1）创建需要启动的 BroadcastReceiver 的 Intent。

　　（2）调用 Context 对象的 sendBroadcast()（发送普通广播）或 sendOrderedBroadcast()（发送有序广播）方法来启动指定的 BroadcastReceiver。

　　BroadcastReceiver 是 Android 四大组件之一，用户开发自己的 BroadcastReceiver 与开发其他组件一样，只需要继承 Android 系统中提供的 BroadcastReceiver 基类，然后实现里面的相关方法即可。通常，只需要实现该类的 onReceive(Context context, Intent intent)抽象方法。在接收到广播后，会立即回调该方法，通过传入的 intent 对象，可以很方便地获取广播传递的一些数据。与其他组件类似，创建好自定义的 BroadcastReceiver 后，并不能马上使用，还需要对其进行注册。与其他组件不同的是，BroadcastReceiver 的注册有两种方式，一种与其他组件类似，在清单文件中通过＜receiver.../＞标签进行注册，也叫静态注册；另一种是在 Java 代码中进行注册，也叫动态注册。例如，对于同一个 BroadcastReceiver 类的子类 MyBroadcastReceiver，两种注册方式如下。

　　在清单文件（AndroidManifest）中注册代码如下。

```
1    <receiver android:name=".MyBroadcastReceiver">
2        <intent-filter>
3            <action android:name="iet.jxufe.cn.android.myBroadcastReceiver"/>
4        </intent-filter>
5    </receiver>
```

　　在 Java 中动态注册代码如下。

```
1    MyBroadcastReceiver myBroadcastReceiver=new MyBroadcastReceiver();
2    IntentFilter filter=new IntentFilter();
3    filter.addAction("iet.jxufe.cn.android.myBroadcastReceiver");
4    registerReceiver(myBroadcastReceiver, filter);
```

　　action 表示广播接收器能接收到的广播的动作，一个广播接收器可以接收多个广播，因此可以添加多个 Action 属性，例如本案例中后台服务中的广播接收器既可以接收前台控制播放/暂停的广播，也可以接收改变音乐播放进度的广播等，代码见 MusicService.java 的第 12～17 行。在本案例中，广播接收器都是通过代码进行动态注册的，当程序运行时进行注册，当退出时取消注册。为了使发送广播时指定的 Action 与广播接收器的 Action 保存一致，避免拼写错误，本案例将所有的 Action 都单独定义成一个常量，当需要引用时使用类名＋常量名，见 Constants.java 文件。

　　注册完成后，即可接收相应的广播消息。一旦广播（Broadcast）事件发生，系统就会创建对应的 BroadcastReceiver 实例,并自动触发它的 onReceive()方法,onReceive()方法执行完后,BroadcastReceiver 的实例就会被销毁。在本案例中,一个广播接收器可以接收多种广播,因此需要判断接收到的究竟是哪一种广播,主要是通过接收到的广播的Action 来确定的。Activity 与 Service 之间的广播传递关系如图 17-11 所示。

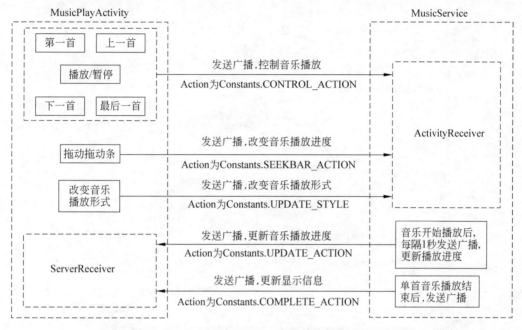

图 17-11　Activity 与 Service 之间的广播传递关系图

17.4　知识扩展

17.4.1　媒体播放器 MediaPlayer

　　Android 的多媒体框架支持一些常见的媒体类型,如音频、视频、图片等,开发者可以很方便地在应用中集成音频、视频等,这些媒体既可以保存在应用程序的资源文件中,也可以保存在手机的内、外存储器中,还可以是来自于网络的数据流。所有这些都需要使用到媒体播放相关 API,其中最简单的就是 MediaPlayer。

　　获取 MediaPlayer 对象有两种方式,一种是调用 MediaPlayer 对象的静态方法 create();另一种是使用 new 关键字来创建。二者的区别在于使用 new 创建的 MediaPlayer 实例处于 Idle 状态,使用 create()方法创建的 MediaPlayer 实例处于 Prepared 状态。关于MediaPlayer 的各种状态之间的关系,官方文档中提供了一个状态转化图,如图 17-12 所示。对各状态介绍如下。

　　• Idle 状态:当使用 new()方法创建一个 MediaPlayer 对象或者调用了 MediaPlayer 对象的 reset()方法时,该 MediaPlayer 对象处于 Idle 状态。若在这个状态下调用 getDuration()等方法,如果是通过 reset()方法进入 Idle 状态,则

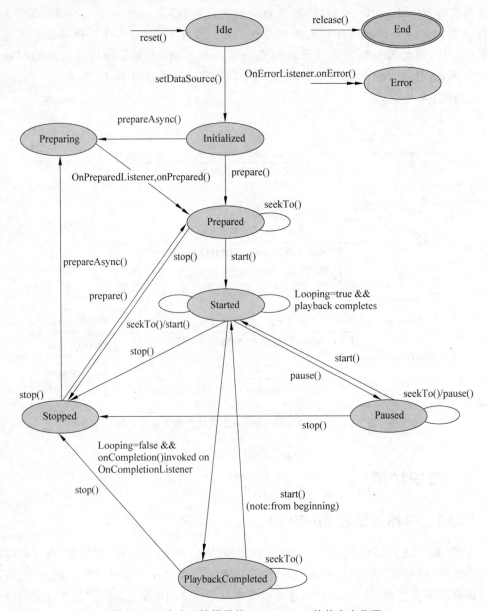

图 17-12 官方文档提供的 MediaPlayer 的状态变化图

会触发 OnErrorListener.onError(),并且 MediaPlayer 会进入 Error 状态;如果是新创建的 MediaPlayer 对象,则不会触发 onError(),也不会进入 Error 状态。

- End 状态:通过 release() 方法可以进入 End 状态,当 MediaPlayer 对象不再被使用时应当尽快将其通过 release() 方法释放,从而释放相关的软/硬件组件资源。如果 MediaPlayer 对象进入了 End 状态,则不会再进入任何其他状态了。
- Initialized 状态:这个状态比较简单,MediaPlayer 调用 setDataSource() 方法后就进入 Initialized 状态,表示此时要播放的文件已经设置好。
- Prepared 状态:初始化完成之后还需要调用 prepare() 或 prepareAsync() 方法,

它们的区别在于prepareAsync()是异步的,它不会阻塞当前的UI线程,只有进入Prepared状态,才表明MediaPlayer到目前为止都没有错误,可以进行文件播放。

- Preparing 状态:这个状态比较好理解,主要是和prepareAsync()配合,如果异步准备完成,会触发OnPreparedListener.onPrepared(),进而进入Prepared状态。
- Started 状态:MediaPlayer一旦准备好,就可以调用start()方法,这样MediaPlayer就处于Started状态,这表明MediaPlayer正在播放文件。用户可以使用isPlaying()测试MediaPlayer是否处于Started状态。如果播放完毕,而又设置了循环播放,则MediaPlayer仍然会处于Started状态。类似地,如果在该状态下MediaPlayer调用了seekTo()或者start()方法,均可以让MediaPlayer停留在Started状态。
- Paused 状态:在Started状态下MediaPlayer调用pause()方法可以暂停MediaPlayer,从而进入Paused状态,MediaPlayer暂停后再次调用start()则可以继续MediaPlayer的播放,转到Started状态,暂停状态时可以调用seekTo()方法,但是不会改变状态。
- Stop 状态:在Started或者Paused状态下均可调用stop()停止MediaPlayer,而处于Stop状态的MediaPlayer要想重新播放,需要通过prepareAsync()或prepare()回到先前的Prepared状态重新开始才可以。
- PlaybackCompleted 状态:文件正常播放完毕,而又没有设置循环播放,则进入该状态,并会触发OnCompletionListener的onCompletion()方法。此时可以调用start()方法重新从头播放文件,也可以调用stop()停止MediaPlayer,或者调用seekTo()重新定位播放位置。
- Error 状态:如果由于某种原因MediaPlayer出现了错误,会触发OnErrorListener.onError()事件,此时MediaPlayer即进入Error状态,及时捕捉并妥善处理这些错误是很重要的,可以帮助我们及时释放相关的软/硬件资源,也可以改善用户体验。通过setOnErrorListener(android.media.MediaPlayer.OnErrorListener)可以设置该监听器。如果MediaPlayer进入了Error状态,可以通过调用reset()来恢复,使得MediaPlayer重新返回到Idle状态。

为了监听一些导致状态变化的事件,MediaPlayer提供了一些绑定事件监听器的方法,最常用的有以下4个。

- setOnCompletionListener(MediaPlayer.OnCompletionListener listener):为MediaPlayer的播放完成事件绑定事件监听器。
- setOnErrorListener(MediaPlayer.OnErrorListener listener):为MediaPlayer的播放错误事件绑定事件监听器。
- setOnPreparedListener(MediaPlayer.OnPreparedListener listener):当MediaPlayer调用prepared()方法时触发该监听器。
- setOnSeekCompleteListener(MediaPlayer.OnSeekCompleteListener listener):当MediaPlayer调用seek()方法时触发该监听器。

本案例主要使用MediaPlayer播放音频,针对不同来源的音频,使用MediaPlayer播放的步骤也不同。具体如下：

(1) 播放应用的资源文件，即音频文件存放在 res/raw 文件夹下。

① 调用 MediaPlayer 的 create(Context context, int resId) 方法加载指定资源文件；

② 调用 MediaPlayer 的 start()、pause()、stop() 等方法控制音频的播放、暂停、停止等。

(2) 播放应用的原始资源文件，即音频文件存放在 assets 目录下。

① 调用 Context 的 getAssets() 方法获取应用的 AssetManager；

② 调用 AssetManager 对象的 openFd(String name) 方法打开指定的原声资源，该方法返回一个 AssetFileDescriptor 对象；

③ 调用 AssetFileDescriptor 的 getFileDescriptor()、getStartOffset() 和 getLength() 方法来获取音频文件的 FileDescriptor、开始位置、长度等；

④ 创建 MediaPlayer 对象，并调用 MediaPlayer 对象的 setDataSource(FileDescriptor fd, long offset, long length) 方法来装载音频资源；

⑤ 调用 MediaPlayer 对象的 prepare() 方法准备音频；

⑥ 调用 MediaPlayer 的 start()、pause()、stop() 等方法控制播放。

(3) 播放手机内、外存储器上的音频文件，即音频文件存放在手机上，与应用软件无关。

① 创建 MediaPlayer 对象，并调用 MediaPlayer 对象的 setDataSource(String path) 方法装载指定的音频文件；

② 调用 MediaPlayer 对象的 prepare() 方法准备音频；

③ 调用 MediaPlayer 的 start()、pause()、stop() 等方法控制音频播放。

(4) 播放来自网络的音频文件，即音频文件存放在网络上。

① 根据网络上的音频文件所在的位置创建 Uri 对象；

② 创建 MediaPlayer 对象，并调用 MediaPlayer 对象的 setDataSource(Context context, Uri uri) 方法装载 Uri 对应的音频文件；

③ 调用 MediaPlayer 对象的 prepare() 方法准备音频；

④ 调用 MediaPlayer 的 start()、pause()、stop() 等方法控制音频播放。

本案例中播放的音乐来自于手机的存储卡，所以使用的是第 3 种，根据音频的位置来播放。

17.4.2 发送通知 Notification

Notification 是显示在手机状态栏上的通知，状态栏位于手机屏幕的最上方，主要显示手机当前的网络状态、电池状态、时间等。Notification 代表的是一种具有全局效果的通知，通过它可以查看该通知的详细信息，Notification 通过 NotificationManager 服务发送。与对话框类似，Notification 的创建也需要借助其内部类 Builder。该类的主要方法如下：

- setAutoCancel()：设置单击通知后状态栏是否自动删除通知，本案例中设置不删除。

- setDefaults()：使用系统的默认设置。
- setContent()：设置通知显示的 View。
- setContentIntent()：设置单击通知后将要启动的程序对应的 PendingIntent。
- setContentText()：设置通知内容。
- setContentTitle()：设置通知标题。
- setLargeIcon()：设置通知的大图标。
- setLights()：设置通知的灯光。
- setSound()：设置通知的声音。
- build()：返回一个已构建好的 Notification。

Notification 通知的主要构成如图 17-13 所示。

①：通知大图标　②：通知标题　③：通知内容
④：通知小图标　⑤：通知时间

图 17-13　Notification 通知的结构图

发送 Notification 的主要步骤如下：

（1）调用 getSystemService() 方法获取系统的 NotificationManager 服务。

（2）通过 Builder 构造器创建一个 Notification 对象。

（3）为 Notification 设置各种属性。

（4）通过 NotificationManager 发送 Notification。

本案例中发送通知的详细代码见 MusicService 的第 **98～111** 行。

17.5　思考与练习

（1）本案例中只能将音乐设置为手机铃声，请在此基础上添加可设置为通知铃声、闹钟铃声功能。

（2）ContentProvider 是 Android 中的四大组件之一，写好 ContentProvider 后需要在清单文件中进行配置，在配置<provider>标签时，以下（　　）属性是必需的。

　　A）android：name　　　　　　　B）android：authorities
　　C）android：exported　　　　　D）A 和 B

（3）ContentProvider 的作用是共享数据，暴露可供操作的接口，其他应用则通过（　　）来操作 ContentProvider 所暴露的数据。

　　A）ContentValues　　　　　　B）ContentResolver
　　C）URI　　　　　　　　　　　D）Context

（4）如果一个应用通过 ContentProvider 共享数据，那么其他应用都可以对该数据进行操作，在读取数据时，可以使用 query 方法，该 query 方法是通过（　　）对象调用的。

　　A）ContentResolver　　　　　B）ContentProvider

C) SQLiteDatabase D) SQLiteHelper

(5) 下列不属于 Service 生命周期的回调方法的是(　　)。

A) onCreate() B) onBind()

C) onStart() D) onStop()

(6) 在开发 Service 组件时,需开发一个类使其继承系统提供的 Service 类,该类中必须实现 Service 中的(　　)方法。

A) onCreate() B) onBind()

C) onStartCommand() D) onUnbind()

(7) 以下关于通过 startService()与 bindService()运行 Service 的说法不正确的是(　　)。

A) startService()运行的 Service 启动后与访问者没有关联,而 bindService()运行的 Service 将与访问者共存亡

B) startService()运行的 Service 将回调 onStartCommand()方法,而 bindService 运行的 Service 将回调 onBind()方法

C) startService()运行的 Service 无法与访问者进行通信、数据传递,bindService()运行的 Service 可在访问者与 Service 之间进行通信、数据传递

D) bindService 运行的 Service 必须实现 onBind()方法,而 startService()运行的 Service 没有这个要求

(8) 关于广播接收器,以下描述正确的是(　　)。

A) 广播接收器只能在清单文件中注册

B) 广播接收器注册后不能注销

C) 广播接收器只能接收自定义的广播消息

D) 广播接收器可以在 Activity 中单独注册与注销

(9) MediaPlayer 在播放音/视频资源之前,需要调用(　　)方法完成准备工作。

A) setDataSource() B) prepare()

C) begin() D) ready()

(10) 使用 MediaPlayer 播放存储在外部存储卡(sdcard)中的音乐文件的操作步骤是(　　)。

A) 使用 MediaPlayer.create()方法传入文件路径返回 MediaPlayer 对象,然后准备播放

B) 直接将音乐文件的路径传入 MediaPlayer 的构造方法中,然后准备播放

C) 先创建 MediaPlayer 对象,然后调用它的 setDataSource 方法设置文件源,再准备播放

D) A 和 C 都可以

第18章

Android 中常见的错误与程序调试方法

Android 是基于 Java 语言的,因此一些简单的语法错误在编译时会自动提示,开发者根据提示信息就能很快地修正。然而编译时正常,并不能表示程序能够正常运行,在运行时可能会出现运行时异常导致程序强制退出,还有一种隐蔽性错误,即程序能够正常运行,但结果却和我们期望的不一致,也就是所谓的逻辑错误。下面主要针对后两种情况的解决方案进行简单介绍。

18.1 程序调试工具

18.1.1 LogCat 工具介绍

在 Android 中为开发者提供了一个记录日志的 Log 类,使用 Log 类可以在程序代码中加入一些"记录点",并可以通过 Eclipse 中的 LogCat 工具查看记录。当程序每次执行到"记录点"时,相应的"记录点"就会在 LogCat 中输出一条信息。开发者通过分析这些记录,就可以检查程序执行的过程是否与我们期望的相符合,并依此判断程序代码中可能出错的区域,以便准确定位。

在默认的 Eclipse 编辑窗口中并没有显示提供的 LogCat 工具,需要开发者从 Eclipse 的窗口中调出来,具体操作为在 Eclipse 菜单中选择 **Windows→Show View→Other→Android→LogCat**,此时在控制台窗口中将出现 LogCat 工具。LogCat 工具的各部分含义如图 18-1 所示。

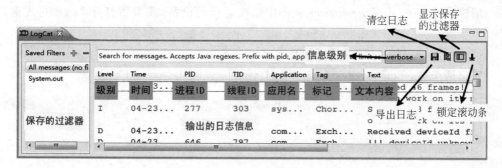

图 18-1 LogCat 工具的各部分含义

在默认情况下,LogCat 中显示的信息比较多,为了显示自己所需要的信息,可以**对信息进行过滤**。添加过滤条件的操作如图 18-2 和图 18-3 所示。

图 18-2 过滤器操作面板

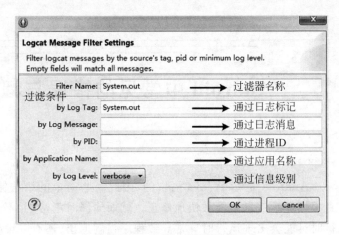

图 18-3 添加过滤器的面板

android.util.Log 中常见的方法有 Log.v()、Log.d()、Log.i()、Log.w() 和 Log.e()，根据首字母分别对应于 VERBOSE、DEBUG、INFO、WARN、ERROR。信息内容从 **ERROR，WARN，INFO，DEBUG，VERBOSE** 依次递增，即 VERBOSE 包含所有的信息，DEBUG 包含 ERROR、WARN、INFO、DEBUG 等信息，而 ERROR 仅仅包含 ERROR 级别的信息。不同类型的信息在 LogCat 中显示的颜色会有所不同，具体如表 18-1 所示。

表 18-1 信息级别及对应颜色表

方法	颜色	消息	方法	颜色	消息
Log.v()	黑	任何信息 verbose	Log.w()	橙	警告信息 warning
Log.d()	蓝	调试信息 debug	Log.e()	红	错误信息 error
Log.i()	绿	提示信息 information			

通常，Log 类中相关的方法需要传递两个参数，一个是信息的标记，即 Tag；另一个是信息的内容。我们可以通过 Tag 标记进行过滤，快速地定位到日志信息。

注意：有时，LogCat 中会不显示任何信息。

解决方法：在 DDMS→devices 视图中选择运行的设备，或重新打开 LogCat，或重启 Eclipse。

简单示例（控制台打印的日志信息顺序）：

定义两个类：Person.java 和 Student.java。

Person.java

```
1   import android.util.Log;
2   public class Person {
3       public Person(){                          //构造方法
4           Log.i(MainActivity.TAG, "Person Construtor invoked!");
5       }
6       public void say(){                        //自定义方法
```

```
7        Log.i(MainActivity.TAG,"Person say() invoked!");
8        System.out.println("I'm a super class!");
9    }
10 }
```

Student.java

```
1  public class Student extends Person {
2      private String name;
3      public Student(){                    //构造方法
4          this("姓名未知");
5          Log.i(MainActivity.TAG, "Student Constructor without argument 
               invoked!");
6      }
7      public Student(String name){         //带一个参数的构造方法
8          this.name=name;
9          Log.i(MainActivity.TAG, "Student Constructor with a argument 
               invoked!");
10     }
11     public void say(){                   //自定义的方法
12         Log.i(MainActivity.TAG,"Student say() invoked!");
13         System.out.println("I'm a subclass of Person! My name is"+name);
14     }
15 }
```

在 MainActivity 类中定义 TAG 常量，并在 onCreate()方法中调用相应方法。

MainActivity.java

```
1  public class MainActivity extends Activity {
2      public static final String TAG="LogCatInfoTest";
3      protected void onCreate(Bundle savedInstanceState) {
4          super.onCreate(savedInstanceState);
5          setContentView(R.layout.activity_main);
6          Person person=new Student();     //多态,父类引用指向子类对象
7          person.say();                    //调用对象方法
8      }
9  }
```

问控制台中日志信息的输出顺序是什么？（选择可能输出的信息，并对其进行排序）

(1) Person Construtor invoked!

(2) Person say() invoked!

(3) I'm a super class!

(4) Student Constructor without argument invoked!

(5) Student Constructor with a argument invoked!

(6) Student say() invoked!

(7) I'm a subclass of Person! My name is XXX

结果：(1)→(5)→(4)→(6)→(7)

18.1.2　Eclipse 提供的 Debug 功能

和 Java 编程一样，在 Eclipse 中也可以对 Android 程序进行调试。首先**在代码中设置断点**，当程序执行到断点时将会停下来。设置断点的方法有以下几种：

(1) 双击左边代码所在行的行号，生成断点标志。

(2) 将光标放在代码所在行，然后右击，选择第一个 Toggle Breakpoint，生成断点标志。

(3) 将光标放在需添加断点的行，然后按 Ctrl＋Shift＋B 组合键，即可生成断点标志。

如果想取消相应的断点，只需重复以上的操作即可。

在设置好断点后，运行程序，此时不再是选择 Run As 而是选择 Debug As。程序会执行到断点处停止，并且跳转到 Debug 视图，如图 18-4 所示。

图 18-4　在 Eclipse 中调试

接下来即可通过调试按钮或快捷键跟踪程序执行的过程，Debug 调试的一些快捷键如下。

- F11：启动 Debug。
- F5：Step into(进入内部执行)。
- F6：Step over(执行下一步)。
- F7：Step Retrun(返回)。
- F8：执行到最后。

18.2　运行时常见的错误

18.2.1　空指针异常

(1) 引用类型的变量只有声明、定义，没有初始化，默认值为 null。

```
1  public class MainActivity extends Activity{
```

第 18 章　Android 中常见的错误与程序调试方法

```
2       private Button login;
3       protected void onCreate(Bundle savedInstanceState){
4           super.onCreate(savedInstanceState);
5           setContentView(R.layout.activity_main);
6           login.setOnClickListener(new OnClickListener(){
7               public void onClick(View v){
8                   System.out.println("登录按钮被单击了!");
9               }
10          });
11      }
12  }
```

此时，编译没有任何错误，但运行时会**抛出空指针异常**！因为并没有具体为 login 赋值，它默认为 null。程序运行结果如图 18-5 或图 18-6 所示。

图 18-5　中文状态下强制退出提示　　　图 18-6　英文状态下强制退出提示

此时就需要查看控制台中对错误信息的描述，一般来说，首先查看错误的开始，对错误的描述，例如：

```
FATAL EXCEPTION: main
java.lang.RuntimeException: Unable to start activit
y ComponentInfo{iet.jxufe.cn.android/iet.jxufe.cn.a
ndroid.MainActivity}: java.lang.NullPointerExceptio
n
```

然后查找 **Caused by** 语句，查看是由什么造成的。

```
Caused by: java.lang.NullPointerException
at iet.jxufe.cn.android.MainActivity.onCreate(Main
Activity.java:14)
```

发现原因后需要对其进行分析，为什么会为 null 值，从而进行相应的修改。
下面的修改行不行呢？为什么？

```
1   public class MainActivity extends Activity{
2       private Button login=(Button)findViewById(R.id.login);
3       protected void onCreate(Bundle savedInstanceState){
4           super.onCreate(savedInstanceState);
5           setContentView(R.layout.activity_main);
6
7           login.setOnClickListener(new OnClickListener(){
8               public void onClick(View v){
9                   System.out.println("登录按钮被单击了!");
```

```
10          }
11      });
12  }
13 }
```

此时，系统仍然会抛出空指针异常，这是因为 findViewById()方法的作用是通过 ID 从某个布局文件中查找相应的控件，它的前提是该布局文件已加载。布局文件的加载是在 onCreate()方法中，login 作为成员变量，是在类加载的时候就执行的，而 onCreate()方法是在创建了该类的对象后才会执行。正确方法如下。

```
1  public class MainActivity extends Activity {
2      private Button login;
3      protected void onCreate(Bundle savedInstanceState){
4          super.onCreate(savedInstanceState);
5          setContentView(R.layout.activity_main);
6          login=(Button)findViewById(R.id.login);
7          login.setOnClickListener(new OnClickListener(){
8              public void onClick(View v){
9                  System.out.println("登录按钮被单击了！");
10             }
11         });
12     }
13 }
```

（2）根据 findViewById()方法未能找到相应控件（主要针对多个布局文件）。

```
1  public class MainActivity extends Activity {
2      private Button login;
3      private Button reset;
4      private EditText name,psd;
5      protected void onCreate(Bundle savedInstanceState){
6          super.onCreate(savedInstanceState);
7          setContentView(R.layout.activity_main);
8          login=(Button)findViewById(R.id.login);
9          login.setOnClickListener(new OnClickListener(){
10             public void onClick(View v){
11                 Builder builder=new AlertDialog.Builder(MainActivity.this);
12                 builder.setTitle("欢迎登录");
13                 View view=getLayoutInflater().inflate(R.layout.login, null);
14                 reset=(Button)findViewById(R.id.reset);
15                 name=(EditText)findViewById(R.id.name);
16                 psd=(EditText)findViewById(R.id.psd);
17                 reset.setOnClickListener(new OnClickListener(){
18                     public void onClick(View v){
```

```
19                    name.setText("");
20                    psd.setText("");
21                }
22            });
23            builder.setView(view);
24            builder.create().show();
25        }
26    });
27 }
28 }
```

默认情况下，Activity 的 findViewById()方法会从 setContentView 方法设置的布局文件中查找控件，但上面的代码中这些控件并不是在 R. layout. activity_main 中，而是在 R. layout. login 中。在上面代码中已经将 R. layout. login 转换成了 View 对象，此时应调用 View 类的 findViewById()。因此，只需将上面加粗的部分用下面的代码替换即可。

```
reset=(Button)view.findViewById(R.id.reset);
name=(EditText)view.findViewById(R.id.name);
psd=(EditText)view.findViewById(R.id.psd);
```

18.2.2 类型转换异常

Android 中 Activity 类的 findViewById()方法的返回值为 View 类型，在实际应用中我们经常需要调用具体控件的一些特殊方法，例如 ImageView 设置图片、TextView 设置文本内容等。而 View 类并没有提供相关的方法，因此需要把 View 对象转换成具体的子类对象。由父类对象强制转换为子类对象，在编译时是不会出错的，但是当程序运行时，如果具体的对象与你所转换的对象类型不一致，也不存在父子关系时，则会抛出类型转换异常。例如，将 ImageView 强制转换成 TextView，将 TextView 转换为 Button 等。而将 Button 强制转换成 TextView 不会出错，因为 TextView 是 Button 的父类，子类对象可以赋给父类引用。

18.2.3 数组越界异常

数组越界异常也是大家在开发中经常会遇到的异常，访问时数组的下标从 0 开始，因此最大的下标为数组的长度减 1，如果访问的下标不在这个范围之内，则抛出数组越界异常。例如循环浏览图片时，当访问到最后一张，如果继续递增，则会导致数组越界。对于数组越界一个比较好的处理方式是将数组的下标设置为当前访问的数对数组的长度取模，这样结果一定在 0 和数组长度－1 之间，不会越界。

18.2.4 重复运行程序出现警告

当前程序已经在前台运行，并且程序没有任何更新，此时重复运行会提示以下警告。

```
Warning: Activity not started, its current task has been brought to the front.
```

即 Activity 没有启动,因为当前任务已经在前台运行。

解决方案:①退出程序再运行;②修改程序再运行,例如添加一个空格。

18.2.5　XML 文件中标签拼写错误

在 Android 开发中,用户还会经常遇到 XML 文件中的单词拼写错误,该错误在编译时不会提示。在程序运行时,则会强制退出,并且 LogCat 中会打印出"android. view. InflateException:Binary XML file line #:Error inflating class Xxxx."的信息。

错误原因:

(1) 引用类名问题,即标签的名称写错,这时候系统根据反射机制找不到相应的类。

(2) 如果是自定义标签,那么自定义的类必须实现包含属性的构造方法。

- View(Context context):仅包含 Context 类型参数的构造方法,通过这种方式自定义的控件只能通过 Java 代码来创建。
- View(Context context, AttributeSet attrs):通过这种方式自定义的控件既可以在 Java 代码中创建,也可以在 XML 文件中使用,在 XML 文件中使用时,使用完整的包名＋类名作为标签的名称,如下所示。

```
1  public class MyButton extends Button {
2      public MyButton(Context context) {
3          super(context);
4      }
5  //  public MyButton(Context context,AttributeSet attrs) {
6  //      super(context,attrs);
7  //  }
8  }
```

18.2.6　使用 ListActivity 时调用 setContentView()方法出错

当使用 ListActivity 时,可以不包含任何布局文件,即不调用 setContentView()方法,如果使用 setContentView()方法设置显示的界面,则在布局文件中必须包含一个 ListView,并且 ListView 的 ID 为 @android:id/list,否则会抛出运行时异常(Fatal Exception 致命的异常):**Your content must hava a ListView whose id attribute is 'android. R. id. list'**。为什么?

ListActivity has a default layout that consists of a single, full-screen list in the center of the screen. However, if you desire, you can customize the screen layout by setting your own view layout with setContentView() in onCreate(). To do this, your own view **MUST** contain a ListView object with the id "@android:id/list"。因为 ListActivity 中有一个默认的布局文件,该文件中仅包含一个占满整个屏幕的 ListView,并且该 ListView 的 ID 为@android:id/list,系统会根据这个 ID 来获取 ListActivity 中的 ListView。

18.2.7 在 Eclipse 中导入项目时错误

1. 几乎所有的 Java 类都报错

出现这种现象,通常是由 Android 的版本造成的,原来项目所使用的版本在本机上不存在,此时我们可以看到在项目的文档结构中不存在 Android 开发包。

解决方案:为该项目引入 Android 开发包,不一定要和原版本一致,可以引入比原版本更高的开发包。操作过程为选中该项目,右击→选择 Properties,弹出对话框→选择 Android,然后在右边选择一个已有的 Android 开发包→Apply→OK。

2. 提示 Java 编译器错误

在 Eclipse 中导入 Android 项目时经常出现:Android requires compiler compliance level 5.0 or 6.0. Found'1.4' instead. Please use Android Tools>Fix Project Properties.

解决方案:

(1) 按提示在工程文件上右击→Android Tools→Fix Project Properties 即可;

(2) 若(1)无效,则手动打开 Project→Properties→java Compiler→选择 Enable project specific setting→再选择 Compiler Compliance Level(选择任意一个非默认的值)→OK;

(3) 重复第(2)步,将 Compiler Compliance Leave 选为正确的值(该值一般是当前安装的 JDK 版本值,如 jdk 5 对应 1.5,jdk 6 对应 1.6)→OK。

3. 提示 @Override 报错

在导入 Android 工程的时候,有时明明是刚刚用过的没有问题的工程,但重新导入的时候就报错。

提示 The method … must override a spuerclass method,然后 Eclipse 给出提示让我们把 @Override 删除。

这个错误源于 **java compiler**,在 Java 1.5 中是没有 @Override 的,在 Java 1.6 中才有。

解决方案:让 Eclipse 使用 Java 1.6 而不是 Java 1.5。

操作过程为在 Eclipse 中选择 Window→Preferences→Java→Compiler。

虽然这个时候我们可能在右边看到的 Compiler compiance level 选择的是 1.6,但是细分到每个项目的时候不一定,因此继续选择"Configure Project Specific Setings…",于是我们可以看到我们的工程了,选择报错的工程→OK。

这时我们看到这里的 JDK Compliance 并不是 1.6,将其修改为 1.6→OK。

第19章

Android 程序员猎头系统

19.1 系统功能概述

随着移动互联网的迅猛发展、智能手机的普及,各行各业纷纷进军移动市场,开发手机客户端或开展移动商务应用,以抢占先机,一时间,移动软件开发人员的需求非常多。而移动互联网的兴起只有短短的几年时间,对于中小型企业而言,期望开拓自己的移动业务,但难以找到合适的人选。如果自己培养一名移动开发人员,则成本较高,而通过社会招聘,风险又比较大,新录用人员的技能水平究竟如何,难以评测。对于应聘者而言,现实中的应聘时间、精力、交通成本较高,同时对自己掌握的技能和企业的需求并不是十分明确。

基于此,本着服务企业与求职者的理念,充分利用自身的优势,我们设计和开发了一套 Android 程序员猎头系统。该系统集求职、测试、评审于一体,既方便企业发布招聘信息、测试应聘人员、降低宣传成本,也利于求职者投递简历、了解岗位需求、随时检测自己的能力水平、降低应聘成本,同时提供相关的学习资源供应聘者学习。

借助 Android 程序员猎头系统,中小型企业可根据自己的需求随时发布招聘信息,同时可以根据不同岗位的要求设置相应的测试题,如果对移动开发不是很熟,也可以让系统随机出题。应聘者通过该系统可以及时了解各企业的招聘需求、投递简历、进行客观题和主观题的在线测试。对于客观题,系统会自动阅卷,及时给出测试成绩,对于主观题,系统会随机分配给至少两位评审专家进行评审。当所有评审专家评审结束后,系统会生成总的成绩报告单,通过该成绩报告单,企业能够及时了解应聘者的知识技能水平,同时应聘者个人也可以发现自己的能力水平,从而查缺补漏。该系统完整地记录了整个招聘过程,提供一个公平、公开、公正的低成本求职招聘平台。

19.2 系统结构

19.2.1 开发技术

为保证 Android 程序员猎头系统使用便捷,本系统采用 B/S(Browser 浏览器/Server 服务器)的体系结构。系统分为四类重要角色,该系统中应聘者主要功能包括填写简历、职位申请、基础题测试、编程题测试。企业用户主要功能包括职位发布、试卷管理、成绩单管理。评委包括参与评审和管理自己的评审记录。管理员又分为四类,不同类型的管理员具有不同的权限。因此该系统属于较大型的 Web 应用项目,诸如此类项目需要很复杂的表现和逻辑处理,如何提高开发的效率,增强系统的可扩展性和维护性,是开发此项目需要面临的问题。本系统采用了 Struts2＋Hibernate 开源框架,可以很好地解决上述问题,使用分层思想,为 Web 应用的各层都提供了良好的框架整合,最大程度地降低系统各层的耦合,提高了系统整体开发效率。

19.2.2 主页面介绍

登录 Android 程序员猎头系统后,将显示系统的主界面。该界面的层次结构非常清楚,共分为 4 个模块。最上面按功能分为两个部分,左侧为导航模块,右侧为登录和注册模块。主界面的中间又分为两个部分,左侧为动态图片展览区,右侧为滚动的企业招聘信息。最下面是本系统提供的行业资讯、求职指导、友情链接等信息模块。主界面如图 19-1 所示。

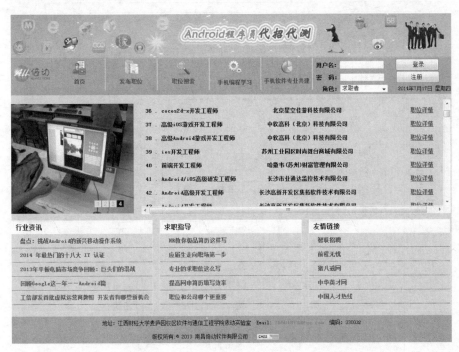

图 19-1　Android 程序员猎头系统主界面

19.2.3 系统功能流程图

Android 程序员猎头系统可以实现网上招聘功能。企业通过本系统发布招聘启事,按岗位设置编程题的试题,用于考核应聘者的技能知识水平。应聘者可以使用浏览器登录本系统了解企业的招聘信息,如果应聘者想通过基础题测试自己对技能的掌握程度,可以重复添加基础题进行测试。当应聘者申请职位时,可以将最好的成绩和简历投递给企业,若企业给予应聘者测试机会,应聘者可以进行编程题的测试。评委查看分配给自己的评审任务并对分配到的试卷进行评审打分和书写评语,最后生成成绩报告单反馈给应聘者和企业。本系统的主要流程如图 19-2 所示。

Android 编程经典案例解析

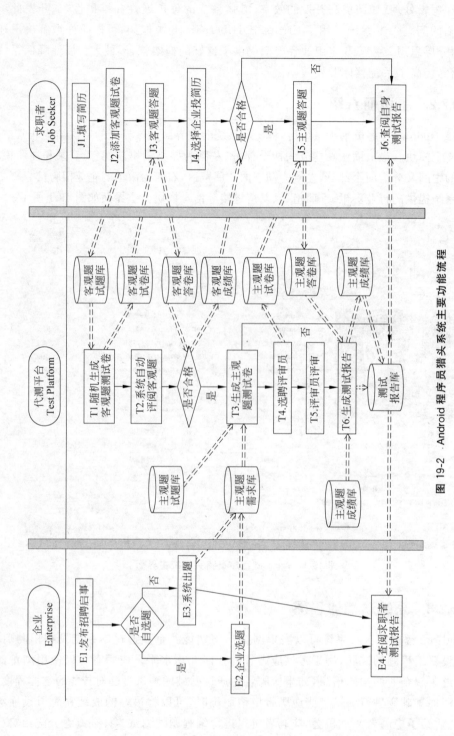

图 19-2 Android 程序员猎头系统主要功能流程

19.3 系统业务操作流程

19.3.1 企业招聘操作流程

本节主要介绍企业招聘的整个操作流程。首先企业需要发布招聘启事，按照岗位设置编程题试卷，接下来是应聘者申请职位和职位测试阶段，最后由评委对应聘者的试卷进行评估。企业通过查看应聘者的简历和总成绩报告单来确定是否录取。企业招聘操作流程如图 19-3 所示。

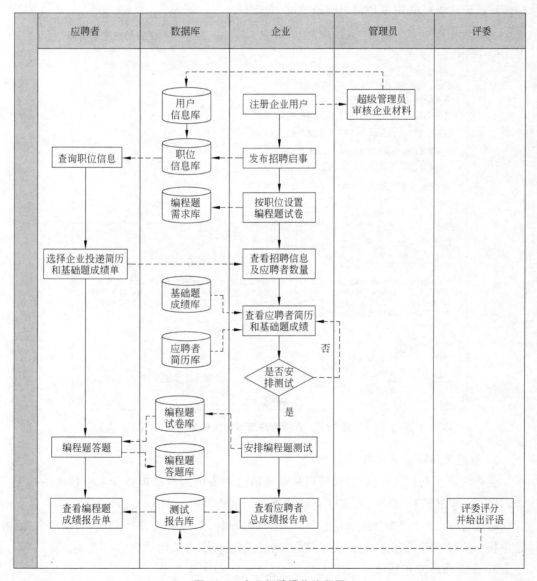

图 19-3　企业招聘操作流程图

1. 企业用户注册

企业用户在使用 Android 程序员猎头系统之前,需要根据自己的角色来注册个人信息。单击导航右侧的"注册"按钮,进入到注册界面。注册界面如图 19-4 所示。

如果注册的用户角色为企业用户或评审,不仅要填写上面的信息,而且单击"下一步"按钮后需要输入用户的审核材料及详细信息。当超级管理员审核通过后,企业用户或评审才能使用已经注册过的用户和密码登录 Android 程序员猎头系统。企业用户填写审核信息如图 19-5 所示。

图 19-4　企业用户注册界面

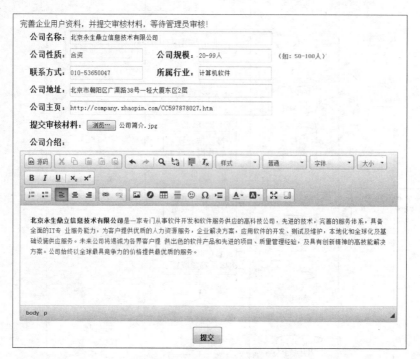

图 19-5　企业用户填写审核信息

2. 管理员审核企业资料

企业注册完成后,超级管理员可以从审核信息列表中查看已审核和未审核的用户记录,其中有"审核资料"信息,可提供下载和查看。审核界面如图 19-6 所示。

超级管理员单击"查看注册信息"浏览用户的资料,如果资料符合要求,可以单击"通过审核",此时审核状态更改为"已审核"。审核界面如图 19-7 所示。

3. 企业用户发布职位

企业用户资料被审核通过后,登录系统单击企业用户管理界面左侧的"发布招聘信息"进入发布招聘信息页面,可以添加需要发布招聘的详细信息,如图 19-8 所示。

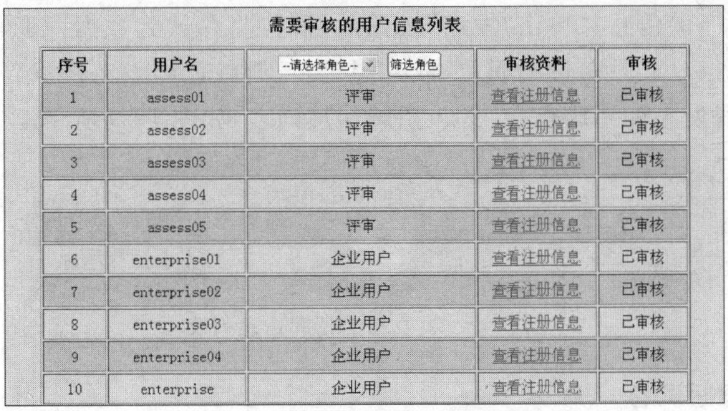

需要审核的用户信息列表

序号	用户名	--请选择角色-- 筛选角色	审核资料	审核
1	assess01	评审	查看注册信息	已审核
2	assess02	评审	查看注册信息	已审核
3	assess03	评审	查看注册信息	已审核
4	assess04	评审	查看注册信息	已审核
5	assess05	评审	查看注册信息	已审核
6	enterprise01	企业用户	查看注册信息	已审核
7	enterprise02	企业用户	查看注册信息	已审核
8	enterprise03	企业用户	查看注册信息	已审核
9	enterprise04	企业用户	查看注册信息	已审核
10	enterprise	企业用户	查看注册信息	已审核

图 19-6 审核企业用户和评审用户的资料

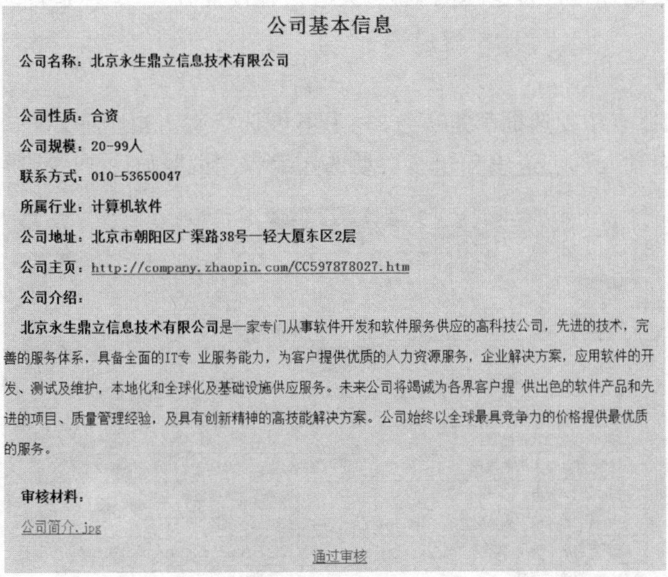

公司基本信息

公司名称：北京永生鼎立信息技术有限公司

公司性质：合资
公司规模：20-99人
联系方式：010-53650047
所属行业：计算机软件
公司地址：北京市朝阳区广渠路38号一轻大厦东区2层
公司主页：http://company.zhaopin.com/CC597878027.htm
公司介绍：
 北京永生鼎立信息技术有限公司是一家专门从事软件开发和软件服务供应的高科技公司，先进的技术，完善的服务体系，具备全面的IT专业服务能力，为客户提供优质的人力资源服务，企业解决方案，应用软件的开发、测试及维护，本地化和全球化及基础设施供应服务。未来公司将竭诚为各界客户提 供出色的软件产品和先进的项目、质量管理经验，及具有创新精神的高技能解决方案。公司始终以全球最具竞争力的价格提供最优质的服务。

审核材料：
公司简介.jpg

通过审核

图 19-7 审核界面

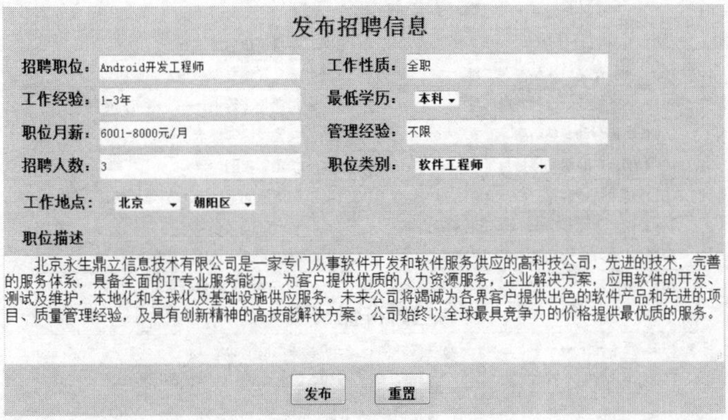

发布招聘信息

招聘职位：Android开发工程师　　工作性质：全职
工作经验：1-3年　　　　　　　　最低学历：本科
职位月薪：6001-8000元/月　　　管理经验：不限
招聘人数：3　　　　　　　　　　职位类别：软件工程师
工作地点：北京　朝阳区
职位描述：
 北京永生鼎立信息技术有限公司是一家专门从事软件开发和软件服务供应的高科技公司，先进的技术，完善的服务体系，具备全面的IT专业服务能力，为客户提供优质的人力资源服务，企业解决方案，应用软件的开发、测试及维护，本地化和全球化及基础设施供应服务。未来公司将竭诚为各界客户提供出色的软件产品和先进的项目、质量管理经验，及具有创新精神的高技能解决方案。公司始终以全球最具竞争力的价格提供最优质的服务。

发布　重置

图 19-8 发布招聘信息界面

企业填写发布的信息并单击"发布"按钮后,职位将存放到数据库职位表中。如果企业用户想查看自己已经发布过的职位,或者想更新、删除已经发布的职位,可以单击左侧功能模块中的"我发布的招聘信息"进入招聘信息列表,如图19-9所示。

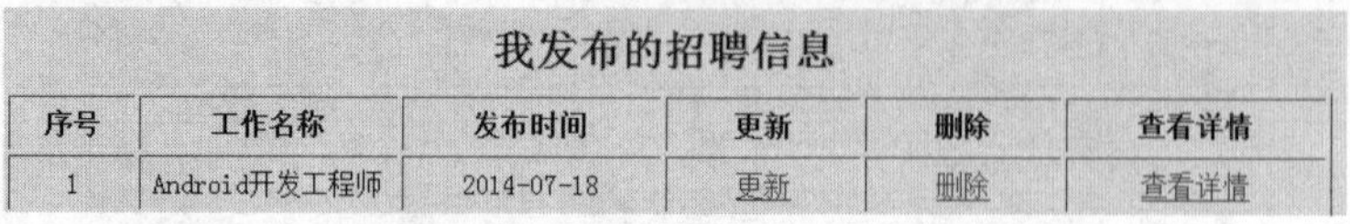

我发布的招聘信息						
序号	工作名称	发布时间	更新	删除	查看详情
1	Android开发工程师	2014-07-18	更新	删除	查看详情

图19-9 我发布的招聘信息列表

4. 设置编程题试卷

企业通过左侧功能模块中的"按岗位设置编程题测试试卷"进入编程题参数设置界面,为自己发布过的职位设置编程测试题,共有两种设置方式,一种是自动选题,另一种是人工选题。

1)自动选题

默认一开始所有的复选框都是灰色的,为不可选状态。当企业用户单击"自动选题"按钮后,由系统自动勾选三道编程题。自动选择编程题试题如图19-10所示。

图19-10 单击"自动选题"系统自动勾选项

2）人工选题

若企业用户不想由系统自动勾选，可以单击"人工选题"按钮，手动选择编程题题目。

5．查看求职者的信息

如果有应聘者申请了企业已经发布的职位，企业用户可以单击左侧功能模块中的"查看应聘者列表"查看申请该职位应聘者的简历和基础题测试成绩报告单。查看招聘信息及求职者数量如图 19-11 所示。

招聘信息及求职者数量			
序号	--请选择职位名称-- ▼ 筛选职位	应聘者数量	查看详情
1	Android开发工程师	1	查看详情

图 19-11　招聘信息及求职者数量

企业可以选择发布的某一职位查看应聘者的详细信息，包括应聘者申请的工作、应聘者的姓名、基础成绩以及应聘者的简历等信息。查看应聘者信息如图 19-12 所示。

应聘者信息					
序号	申请工作	申请者	基础成绩	编程题成绩	报告单	查看简历
1	Android开发工程师	江小璐	查看成绩	未安排测试	未安排测试	查看简历

图 19-12　应聘者详细信息

企业查看应聘者的基础题成绩，可以单击基础成绩下面的"查看成绩"进行查看。查看基础题成绩报告单如图 19-13 所示。

客观题成绩报告单					
考生姓名：江小璐　开始时间：2014-07-29 10:04:26　结束时间：2014-07-29 10:08:40					
序号	知识点	分值小计	得分小计	得分比	掌握程度
1	Java	4	0	0%	不及格
	Android	13	13	100%	优
合计	综合	17	13	76%	中

图 19-13　企业查看应聘者的基础成绩报告单

企业查看应聘者的简历，可以单击查看简历下面的"查看简历"进行查看。企业查看应聘者的简历如图 19-14 所示。

企业可以安排应聘者进行编程题的测试，单击"安排测试"按钮，应聘者可以进入编程题的测试界面。

6．查看应聘者成绩报告单

在应聘者测试完编程题，并提交编程题的试卷后，将等待评审的评分。评审任务由评审业务管理员进行分配，当评审完毕后，将成绩单反馈给应聘者和企业用户查看。查看应聘者信息如图 19-15 所示。

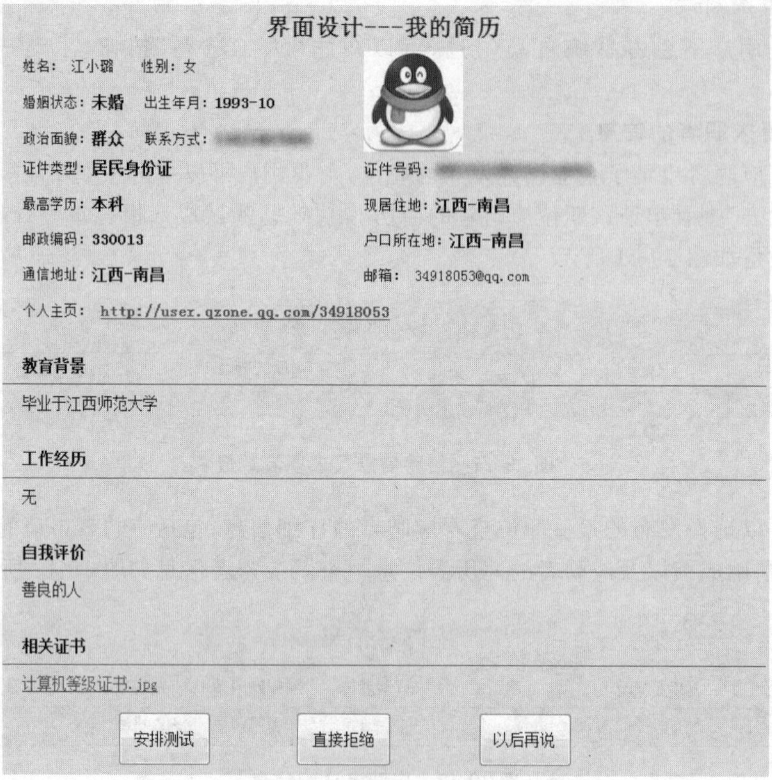

图 19-14 企业查看应聘者简历

应聘者信息

序号	申请工作	申请者	基础成绩	编程题成绩	报告单	查看简历
1	Android开发工程师	江小璐	查看成绩	查看成绩	查看总报告单	查看简历

图 19-15 企业查看应聘者信息列表

企业用户可以单击编程题成绩一栏下面的“查看成绩”查看应聘者的编程题成绩，编程题成绩报告单如图 19-16 所示。

7. 打印应聘者总成绩报告单

若企业用户想打印应聘者的总成绩报告单，可以单击左上角的“打印总成绩单”进行打印，总成绩报告单及打印总成绩报告单如图 19-17 和图 19-18 所示。

19.3.2 应聘者求职操作流程

本节主要介绍企业应聘者求职的操作流程。应聘者可以查询企业发布的招聘计划，当应聘者找到了适合自己的职位时，可以将简历和自测的基础题成绩单投递给企业。如果企业给予编程题测试，应聘者可以参加编程题的测试。应聘者提交的编程题试卷被评委评审完毕后，应聘者可以查看自己的成绩单。应聘者求职操作流程如图 19-19 所示。

编程题成绩报告单

考生姓名：江小璐　开始时间：2014-07-18 09:38:22　结束时间：2014-07-29 10:29:15

序号	编程题题目号	题目名称	分值	得分	得分比
1	T63	T003.我的课表—表格布局应用	30	15.5	52%
	评语A	部分功能实现			
	评语B	部分功能实现			
2	T22	J007.仿QQ好友列表效果	20	14.5	72%
	评语A	功能实现，代码结构较乱			
	评语B	部分功能实现			
3	T20	J005.动态改变按钮图片	20	11.5	57%
	评语A	基本功能实现，界面有待改善			
	评语B	功能基本实现			
合计			70	41	59%

图 19-16　企业查看应聘者编程题成绩报告单

打印总成绩单

考生江小璐——总成绩报告单

客观题成绩					
序号	知识点	分值	得分	得分比	掌握程度
1	Java	4	0	0%	不及格
	Android	13	13	100%	优
合计	综合	17	13	76%	中

编程题成绩					
序号	编程题题目号	题目名称	分值	得分	得分比
1	T20	J005.动态改变按钮图片	20	11.5	57%
	评语A	基本功能实现，界面有待改善			
	评语B	功能基本实现			
2	T63	T003.我的课表—表格布局应用	30	15.5	52%
	评语A	部分功能实现			
	评语B	部分功能实现			
3	T22	J007.仿QQ好友列表效果	20	14.5	72%
	评语A	功能实现，代码结构较乱			
	评语B	部分功能实现			
合计			70	41.5	59%

图 19-17　应聘者总成绩报告单

考生江小璐——总成绩报告单

客观题成绩					
序号	知识点	分值	得分	得分比	掌握程度
1	Java	4	0	0%	不及格
	Android	13	13	100%	优
合计	综合	17	13	76%	中

图 19-18　应聘者总成绩报告单打印界面

Android 编程经典案例解析

编程题成绩					
序号	编程题题目号	题目名称	分值	得分	得分比
1	T20	J005.动态改变按钮图片	20	11.5	57%
	评语A	基本功能实现,界面有待改善			
	评语B	功能基本实现			
2	T63	T003.我的课表—表格布局应用	30	15.5	52%
	评语A	部分功能实现			
	评语B	部分功能实现			
3	T22	J007.仿QQ好友列表效果	20	14.5	72%
	评语A	功能实现,代码结构较乱			
	评语B	部分功能实现			
合计			70	41.5	59%

图 19-18 (续)

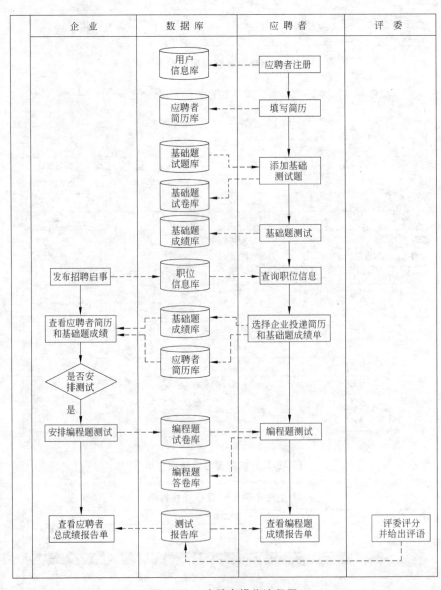

图 19-19 应聘者操作流程图

1．应聘者注册

应聘者在使用 Android 程序员猎头系统之前，需要根据自己的角色来注册个人信息。单击导航右侧的"注册"按钮，进入到注册界面。注册界面如图 19-20 所示。

2．填写简历

应聘者可以在不申请工作的情况下事先把简历填写好，也可以先寻找工作，看见合适的职位再单击申请，在选择简历一栏选择填写新的简历触发本功能。应聘者也可以单击左侧功能列表中的"填写简历"，进入基本信息页面。填写个人简历界面如图 19-21 所示。

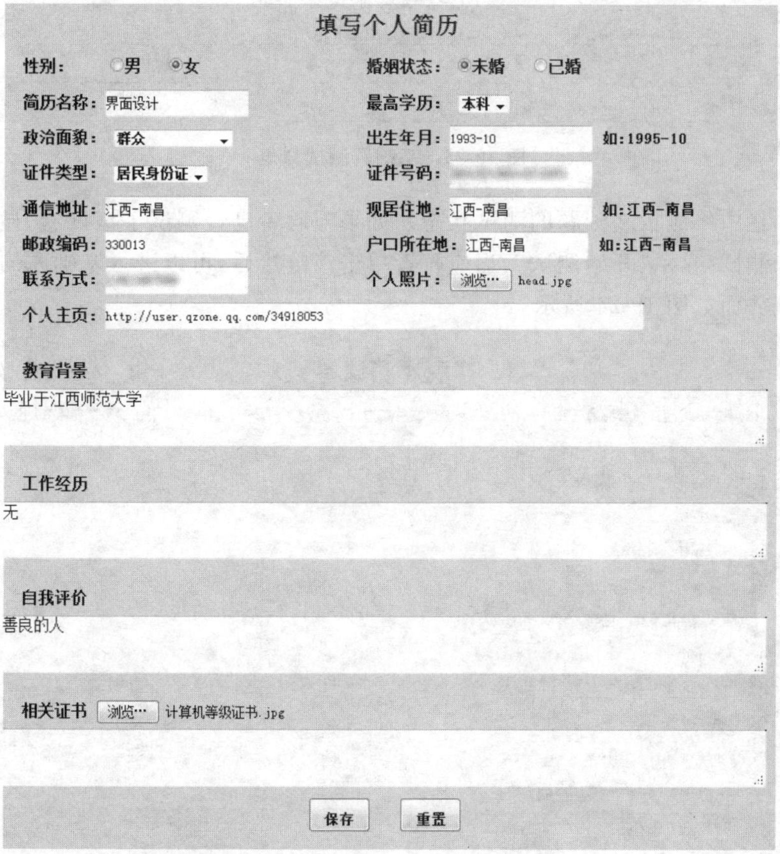

图 19-20　应聘者注册界面

图 19-21　应聘者个人简历页面

3．查看和修改简历

如果应聘者已经填写完个人简历并保存，若想修改简历的内容，可以单击左侧功能模块中的"我的简历列表"，选择"更新"进入更新简历界面。我的简历列表如图 19-22 所示。

我的简历列表

序号	简历编号	简历名称	更新日期	更新	删除	查看详情
1	402881e8477fc46201477fdc2c300003	界面设计	2014-07-29 10:04:01	更新	删除	查看详情

图 19-22　应聘者的简历列表

4. 基础题的测试

应聘者单击"客观题测试"后，进入客观题测试列表界面。当应聘者想进行基础题的测试时，可以单击"添加客观题测试试卷"，此时在列表中将新添加一条测试记录。这里应聘者可以多次添加基础题试卷进行测试。在申请职位时，可以选择最优的成绩投递给企业。基础题测试列表如图 19-23 所示。

客观题测试

添加客观题测试试卷

序号	测试状态	操作
1	待测试	开始测试

图 19-23　基础题测试列表

在操作栏中将显示基础题的测试状态。如果应聘者已经测试了该试卷，显示"查看试卷"；如果未测试，显示"开始测试"，当应聘者单击该链接后，可以进入基础题的测试界面。基础题测试界面如图 19-24 所示。

客观题测试卷

1、Android中的四大组件通常都会在AndroidManifest清单文件中进行注册，以下哪一个组件可以不在清单文件中注册也可以使用。（　　）

　　A、Activity
　　B、Service
　　C、ContentProvider
　　D、BroadcastReceiver

2、为一个boolean类型变量赋值时，可以使用()方式

　　A、boolean = 1;
　　B、boolean a = (9 >= 10);
　　C、boolean a="真";
　　D、boolean a = = false;

3、以下不属于Android中的布局管理器的是（　　）。

　　A、FrameLayout
　　B、GridLayout
　　C、BorderLayout
　　D、TableLayout

4、下列描述中，不正确的是（　　）。

　　A、Android应用的gen目录下的R.java被删除后会自动生成。
　　B、Android项目中res目录是一个特殊目录，用于存放应用中的各种资源，命名规则可以支持大小写字母(a-z，A-Z)、数字(0-9)以及横线(_)。

图 19-24　基础题测试界面

C、AndroidManifest清单文件是每个Android项目必须有的，是项目应用的全局描述，通过包名+组件名可以指定组件的完整路径。

D、Android项目的assets和res目录都能存放资源文件，但是与res不同的是assets支持任意深度的子目录，在它里面的文件不会在R.java里生成任何资源ID。

5、做Android开发时，可以在LogCat视图中查看程序运行时打印的日志信息，如果想测试程序是否执行到某处，只需在该处通过System.out.print()打印一条信息，然后在LogCat中查看是否有该信息即可。

○ F ○ T

6、BroadcastReceiver是Android中的四大组件之一，与其他组件一样使用之前，一定要在清单文件中对其进行注册。

○ F ○ T

7、在类体外不能直接访问类的共有成员

○ F ○ T

8、下面关于Java接口的说法正确的有（ ）

☐ A、一个Java接口是一些方法特征的集合，但没有方法的实现

☐ B、Java接口中定义的方法在不同的地方被实现，可以具有完全不同的行为

☐ C、Java接口中可以声明私有成员

☐ D、Java接口不能被实例化

图 19-24　（续）

5. 查看基础题报告单

应聘者做完客观题的题目后，单击"交卷"，系统将试卷的答案与数据库中的答卷进行匹配，最后生成客观题成绩报告单。该报告单以 Java 和 Android 不同知识点的形式给出得分及得分比例，更能显示求职者对基础知识的掌握程度。

应聘者单击左侧功能模块中的"我的成绩单"进入成绩单列表界面，如图 19-25 所示。

我的成绩单列表						
序号	答卷编号	提交时间	姓名	总分	得分	查看详情
1	23	2014-07-29 10:08:40	江小璐	17	13	查看详情

图 19-25　成绩单列表界面

如果应聘者想查看基础题的成绩单报告，单击"查看详情"，报告单如图 19-26 所示。

客观题成绩报告单					
考生姓名：江小璐　开始时间：2014-07-29 10:04:26　结束时间：2014-07-29 10:08:40					
序号	知识点	分值小计	得分小计	得分比	掌握程度
1	Java	4	0	0%	不及格
	Android	13	13	100%	优
合计	综合	17	13	76%	中

图 19-26　基础题成绩报告单

6. 职位查询

应聘者登录到招聘网站后，可以单击导航中的"搜索职位"，进入职位搜索页面。本系统添加了筛选条件，应聘者可以按照输入的职位名称进行搜索；也可以根据选择的职位类别和工作所在的省份搜索职位。按照软件工程师的工作类型和北京地区的工作地点搜索职位信息如图 19-27 所示。

图 19-27 应聘者查询职位

应聘者可以单击"职位详情"查看职位的详细信息。职位详情如图 19-28 所示。

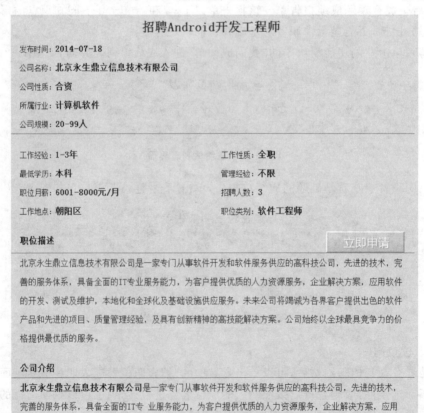

图 19-28 职位详细信息

软件的开发、测试及维护,本地化和全球化及基础设施供应服务。未来公司将竭诚为各界客户提供出色的软件产品和先进的项目、质量管理经验,及具有创新精神的高技能解决方案。公司始终以全球最具竞争力的价格提供最优质的服务。

联系方式: 010-53650047

邮箱: enter20140717@qq.com

公司地址: 北京市朝阳区广渠路38号—轻大厦东区2层

公司主页: http://company.zhaopin.com/CC597878027.htm

图 19-28 （续）

7. 申请职位

单击"立即申请"后,系统将显示申请职位的名称、工作类型、工作地点及公司名称,并且提示应聘者选择简历和客观题成绩报告单。投递简历如图 19-29 所示。

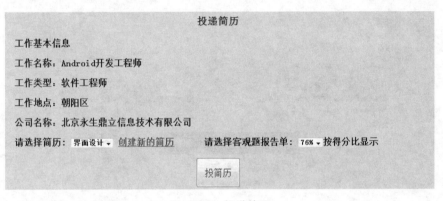

图 19-29 投递简历

8. 查看申请的职位信息

应聘者申请职位后,可以单击"我申请的工作"查看已经申请的职位的详细信息,查看已经申请的工作信息列表如图 19-30 所示。

我申请的工作信息				
序号	简历名称	工作名称	公司名称	查看详情
1	界面设计	Android开发工程师	北京永生鼎立信息技术有限公司	查看详情

图 19-30 申请的职位信息

9. 编程题的测试

应聘者申请职位后,如果企业用户已设置该职位的编程题参数,并安排了应聘者进行测试,当应聘者单击左侧功能模块中的"编程题测试"时会显示可测试的试卷信息。编程题测试列表如图 19-31 所示。

当应聘者单击"开始测试"时,可以进入编程题测试列表界面进行编程题的测试。编程题测试界面如图 19-32 所示。

Android 编程经典案例解析

图 19-31　编程题测试列表

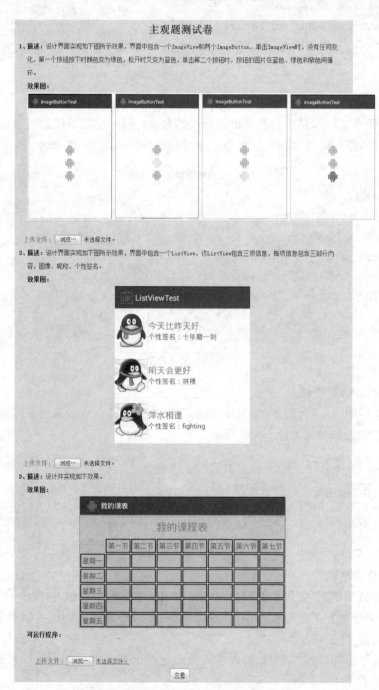

图 19-32　编程题测试界面

10. 查看编程题报告单

在应聘者提交编程题试卷后,等待评委评分;在评委评分完毕后,即可生成编程题报告单进行查看。编程题成绩报告单如图19-33所示。

编程题成绩报告单

考生姓名:江小璐　开始时间:2014-07-18 09:38:22　结束时间:2014-07-29 10:29:15

序号	编程题题目号	题目名称	分值	得分	得分比
1	T22	J007.仿QQ好友列表效果	20	14.5	72%
	评语A	功能实现,代码结构较乱			
	评语B	部分功能实现			
2	T20	J005.动态改变按钮图片	20	11.5	57%
	评语A	基本功能实现,界面有待改善			
	评语B	功能基本实现			
3	T63	T003.我的课表—表格布局应用	30	15.5	52%
	评语A	部分功能实现			
	评语B	部分功能实现			
合计			70	41	59%

图 19-33　编程题报告单

19.3.3　社交化测试流程

在Android程序员猎头系统中社交化测试是核心功能,社交化测试整合了传统的企业招聘模式。企业在发布职位后,可以从测试题库中根据职位选择测试题目,如果应聘者申请了该职位,而且企业安排应聘者测试题时,应聘者将进行编程题的测试。从题目难度上看,保证了职位与测试题目的一致性。应聘者提交已测试完的编程题答卷后,由评审业务管理员根据编程题题目为评委安排评审任务。管理员分配评审任务的模式,能够使评委对应聘者的编程题进行客观准确的评价,从而提高了测试评价体系。如果应聘者的三道题目都被评委评审完毕,系统将自动生成编程题报告单,并将编程题报告单反馈给企业和应聘者。本系统的社交化测试模式可以检测应聘者的专业技能,编程题的成绩报告单将成为企业招聘应聘者的重要参考资料之一。

本节介绍的社交化测试过程按照编程题测试、评审业务管理员分配评审任务、评委评审等环节进行讲述。由于其他节已经多次讲述编程题的测试流程,因此,本节在介绍社交化测试过程时简单叙述应聘者编程题测试流程,重点介绍评审业务管理员的评审任务分配及评审评分过程。社交化测试流程如图19-34所示。

1. 编程题的测试

企业发布职位后,应聘者可以通过Android程序员猎头系统搜索到适合自己的职位。如果应聘者申请职位成功,应聘者可以选择功能模块"我申请的工作"来查看已经申请的职位信息列表。应聘者已经申请的职位信息列表如图19-35所示。

如果北京永生鼎立信息技术有限公司在发布Android开发工程师后,同时设置了该

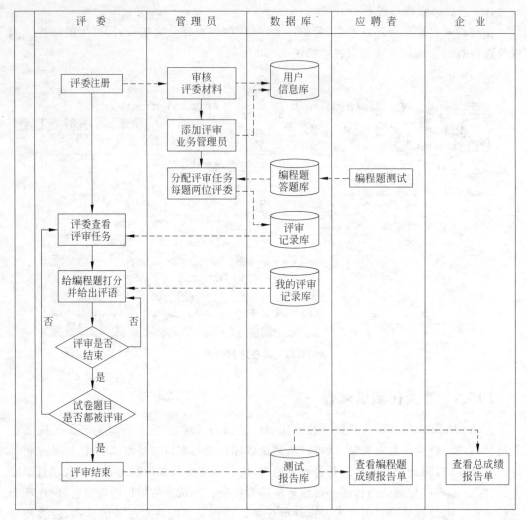

图 19-34 社交化测试流程图

序号	简历名称	工作名称	公司名称	查看详情
		我申请的工作信息		
1	软件	Android开发工程师	北京永生鼎立信息技术有限公司	查看详情

图 19-35 应聘者已经申请的职位信息列表

职位对应的编程题测试试卷，应聘者可以选择模块"编程题测试"进入编程题测试信息列表界面，为自己申请的 Android 开发工程师岗位进行编程题的测试。北京永生鼎立信息技术有限公司的 Android 开发工程师岗位测试信息列表如图 19-36 所示。

单击"开始测试"进入到编程题的测试界面，由于应聘者求职操作流程中讲述得非常清楚，这里不再重复给出编程题的测试界面。如果应聘者提交了已经测试完的编程题答卷，将等待评委的评分。

主观题测试			
序号	工作名称	公司名称	测试状态	操作
1	Android开发工程师	北京永生鼎立信息技术有限公司	待测试	开始测试

图 19-36　岗位测试信息列表

2. 管理员分配评审任务

应聘者提交编程题的答卷后，评审业务管理员可以从评审记录列表中看到应聘者已经提交的编程题题目、答卷编号。他们对应的评审状态如果为未分配状态，则"分配评审员"为可链接状态。评审记录信息列表如图 19-37 所示。

评审记录列表									
序号	答卷编号	题目编号	评审程序	评审A	评审B	评审分数A	评审分数B	最后分数	评审状态	分配评审员
1	3	23	140195301123600150023.rar	assess01	assess02	10	10	10.0	评审结束	已分配
2	3	70	140195301124200150070.rar	assess01	assess02	20	30	25.0	评审结束	已分配
3	3	71	140195301124700150071.rar	assess01	assess02	20	25	22.5	评审结束	已分配
4	6	23	140202359816200160023.rar	assess03	assess04	16	10	13.0	评审结束	已分配
5	6	70	140202359818900160070.rar	assess03	assess04	30	30	30.0	评审结束	已分配
6	6	71	140202359819300160071.rar	assess03	assess04	10	15	12.5	评审结束	已分配
7	9	20	140564960064000010020.zip	assess01	assess02	10	13	11.5	评审结束	已分配
8	9	22	140564960068700010022.zip	assess01	assess02	15	16	15.5	评审结束	已分配
9	9	63	140564960068700010063.zip	assess01	assess02	15	15	15.0	评审结束	已分配
10	11	20	140567327223100170020.zip	assess01	assess02	10	12	11.0	评审结束	已分配
11	11	22	140567327226200170022.zip	assess01	assess02	15	15	15.0	评审结束	已分配
12	11	63	140567327226200170063.zip	assess01	assess02	15	16	14.0	评审结束	已分配
13	13	20	140591035016900020020.zip	assess01	assess02	12	10	11.0	评审结束	已分配
14	13	22	140591035018500020022.zip	assess01	assess02	10	13	11.5	评审结束	已分配

图 19-37　评审记录信息列表

管理员单击"分配评审员"链接后，进入到分配评审员界面，该界面信息表示题号为 20，评审程序为 20140729text1.rar，该题所对应的参考答案为 bc62....alfa.rar，题目总分为 20 的题目分配给 assess01 和 assess02 两位评审员进行评分。如果已经选择好两位评审员，可以单击"分配"按钮完成分配的过程。分配评审员界面如图 19-38 所示。

管理员给题号是 20 的题目分配好两位评审员后，评审记录列表中所对应的评审员姓名将自动填入该条记录中，并且分配评审员的状态改为"已分配"状态，题号为 20 的题目分配评委信息列表如图 19-39 所示。

图 19-38　分配评审员界面

12	24	20	20140729text1.rar	ssess01	assess02	0	0	0.0	未评审	已分配
13	24	22	20140729text2.rar			0	0	0.0	未分配	分配评审员
14	24	63	20140729text3.rar			0	0	0.0	未评审	分配评审员

图 19-39　题号为 20 的题目分配评委信息列表

3. 评委评审

评审业务管理员分配完评审任务后，评委可以查看自己的评审任务，评审任务列表如图 19-40 所示。

图 19-40　评审任务列表

评委可以根据题目编号或者评审状态来搜索评审任务,如果评委在评审状态下输入待评审,则列表中将显示待评审的所有记录。待评审的搜索记录如图 19-41 所示。

序号	题目编号	评审程序	姓名	评审分数	评语	整体状态	我的状态
1	20	20140729text1.rar	assess01	0		未评审	待评审
2	22	20140729text2.rar	assess01	0		未评审	待评审
3	63	20140729text3.rar	assess01	0		未评审	待评审

图 19-41　待评审的搜索记录

评委单击"待评审"后进入评审的界面,给题号为 20、评审程序为 20140729text1.rar、题目所对应参考答案为 bc62....alfa.rar、题目总分为 20 的题目进行评分。界面如图 19-42 所示。

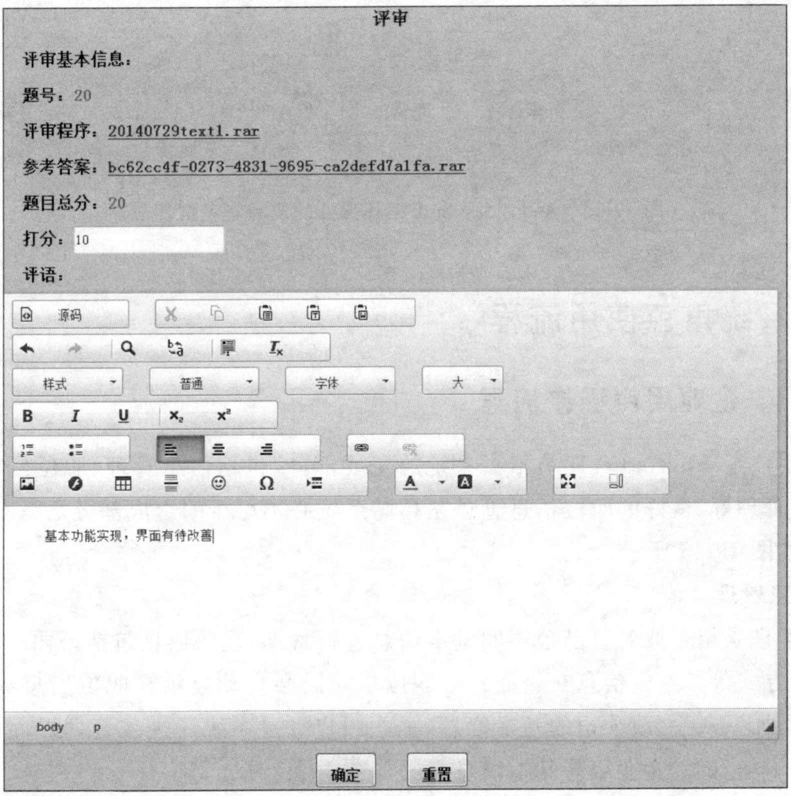

图 19-42　评审界面

评委根据应聘者完成的情况进行打分,并给出该题的评语。单击"确定"按钮,评审任务列表中将会显示该题所对应的分数及评语。A 评委查看 20 号题的评审信息列表如图 19-43 所示。

| 10 | 20 | 20140729text1.rar | assess01 | 10 | 基本功能实现,界面有待改善 | A已评审 | 待评审 |

图 19-43　A 评委查看 20 号题的评审信息列表

A 评委虽然已经评审完 20 号题，但如果 B 评委还没有评审，A 评审可以更改自己的评分和评语。如果 B 评委也已经评审完 20 号题目，则评审任务为评审结束状态。B 评委查看 20 号题目的评审信息列表如图 19-44 所示。

图 19-44　B 评委查看 20 号题的评审信息列表

4. 企业用户查看成绩

如果答卷中所有的题目都已经评审结束，则系统会自动生成编程题成绩报告单反馈给应聘者和企业查看。北京永生鼎立信息技术有限公司可以查看 Android 开发工程师岗位所有应聘者的考试分数。申请该公司 Android 开发工程师岗位的应聘者考试信息如图 19-45 所示。

应聘者信息						
序号	申请工作	申请者	基础成绩	编程题成绩	报告单	查看简历
1	Android开发工程师	江小璐	查看成绩	查看成绩	查看总报告单	查看简历

图 19-45　Android 开发工程师岗位的应聘者考试信息

19.4　系统角色使用流程

19.4.1　企业用户操作流程

企业用户是 Android 程序员猎头系统的主要部分，包括信息管理、职位发布、试卷管理、成绩单管理等，良好的职位信息管理是招聘系统必不可少的组成部分之一。企业用户操作流程如图 19-46 所示。

1. 信息管理

信息管理是指企业对自己公司的基本信息进行管理，这些信息包括公司的介绍、公司地址、联系方式等。这些信息在企业发布职位成功后会自动显示在职位信息界面而不用每次手动添加，若想修改公司的基本信息，企业用户登录系统后进入管理界面，单击左侧的"更新基本信息"。企业填写基本信息界面如图 19-47 所示。

2. 发布职位

单击企业用户管理界面左侧的"发布招聘信息"进入发布招聘信息页面，可以添加需要发布招聘的详细信息。企业发布招聘信息界面如图 19-48 所示。

3. 设置编程题参数

企业通过左侧功能模块中的"按岗位设置编程题测试试卷"进入编程题参数设置界面，为自己发布过的职位设置编程测试题。下面有两种设置方式，一种是自动选题，另一种是人工选题。单击"自动选题"按钮设置编程题测试题界面如图 19-49 所示。

第 19 章　Android 程序员猎头系统

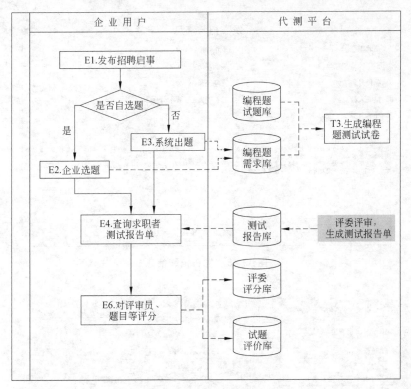

图 19-46　企业用户操作流程图

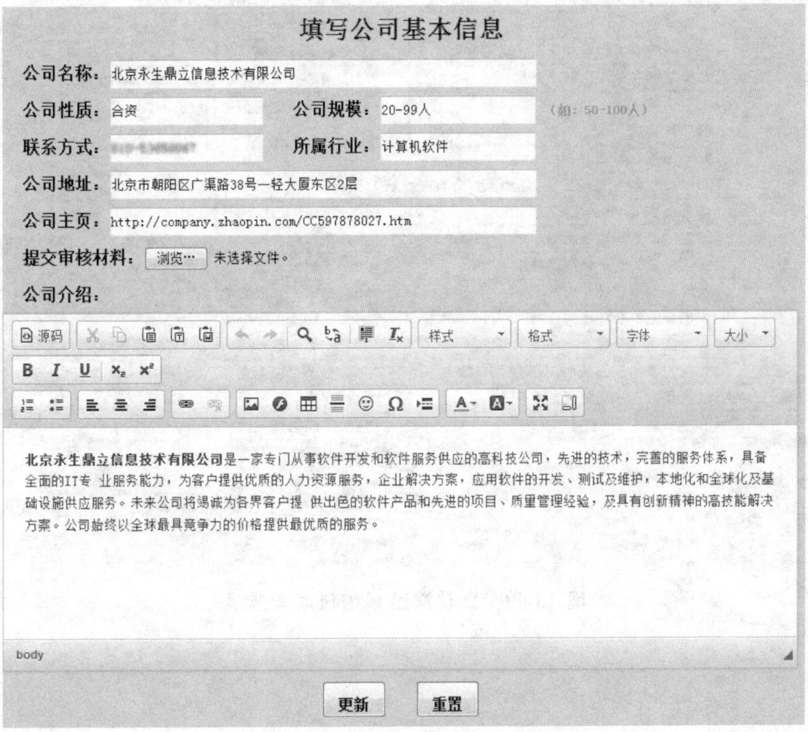

图 19-47　填写公司信息

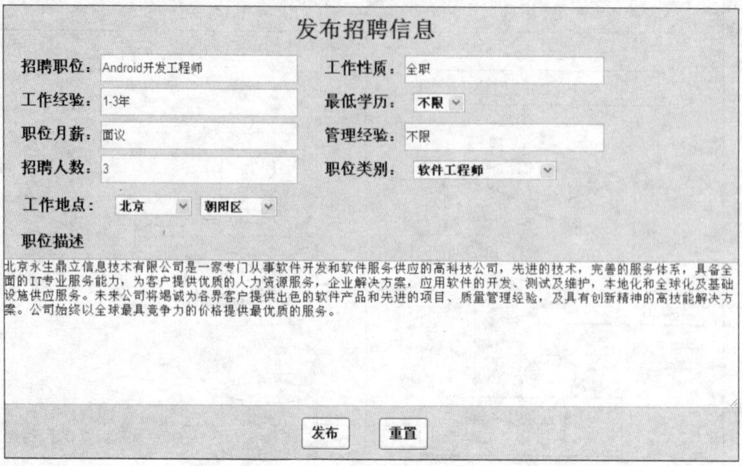

图 19-48　发布招聘信息

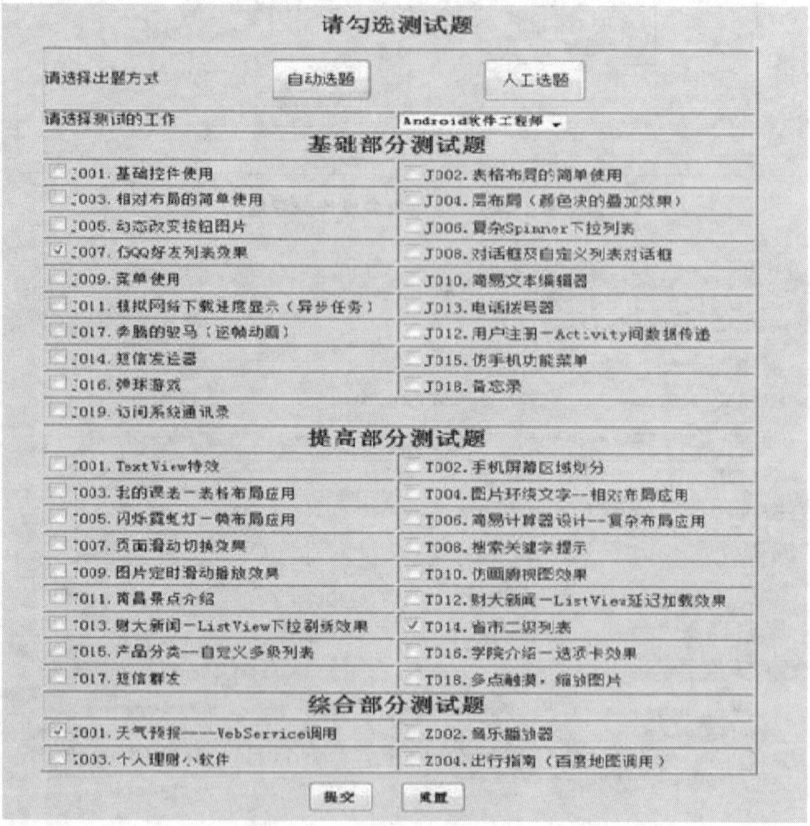

图 19-49　自动选题系统自动勾选项

1）自动选题

默认一开始所有的复选框都是灰色的，为不可选状态。当企业用户单击“自动选题”按钮后，由系统自动勾选三道编程题。

2）人工选题

若企业用户不想由系统自动勾选，可以单击“人工选题”按钮，手动选择编程题题目。

4. 查看应聘者简历和基础题成绩

企业通过左侧功能模块中的“我发布的招聘信息”查看自己发布的职位有哪些应聘者申请过了，以及申请者的数量。企业查看应聘者信息及求职者数量如图 19-50 所示。

招聘信息及求职者数量

筛选职位

序号	--请选择职位名称-- ▾　筛选职位	应聘者数量	查看详情
1	Android软件工程师	1	查看详情

图 19-50　招聘信息及求职者数量

企业可以选择发布的某一职位查看应聘者的详细信息，包括应聘者申请的工作、应聘者的姓名、基础成绩以及应聘者的简历等信息，查看应聘者信息列表如图 19-51 所示。

应聘者信息

序号	申请工作	申请者	基础成绩	编程题成绩	报告单	查看简历
1	Android开发工程师	江小璐	查看成绩	查看成绩	查看总报告单	查看简历

图 19-51　应聘者详细信息

企业查看应聘者的基础题成绩，可以单击“基础成绩”下面的“查看成绩”进行查看，查看应聘者基础题成绩报告单如图 19-52 所示。

客观题成绩报告单

考生姓名：江小璐　　开始时间:2014-07-29 10:04:26　　结束时间:2014-07-29 10:08:40

序号	知识点	分值小计	得分小计	得分比	掌握程度
1	Java	4	0	0%	不及格
	Android	13	13	100%	优
合计	综合	17	13	76%	中

图 19-52　企业查看应聘者的基础成绩报告单

企业查看应聘者的简历，可以单击“查看简历”下面的“查看简历”进行查看，应聘者简历如图 19-53 所示。

5. 查看和打印应聘者总成绩报告单

若企业用户想打印应聘者的总成绩报告单，可以单击左上角的“打印总成绩单”进行打印。企业打印应聘者总成绩报告单如图 19-54 和图 19-55 所示。

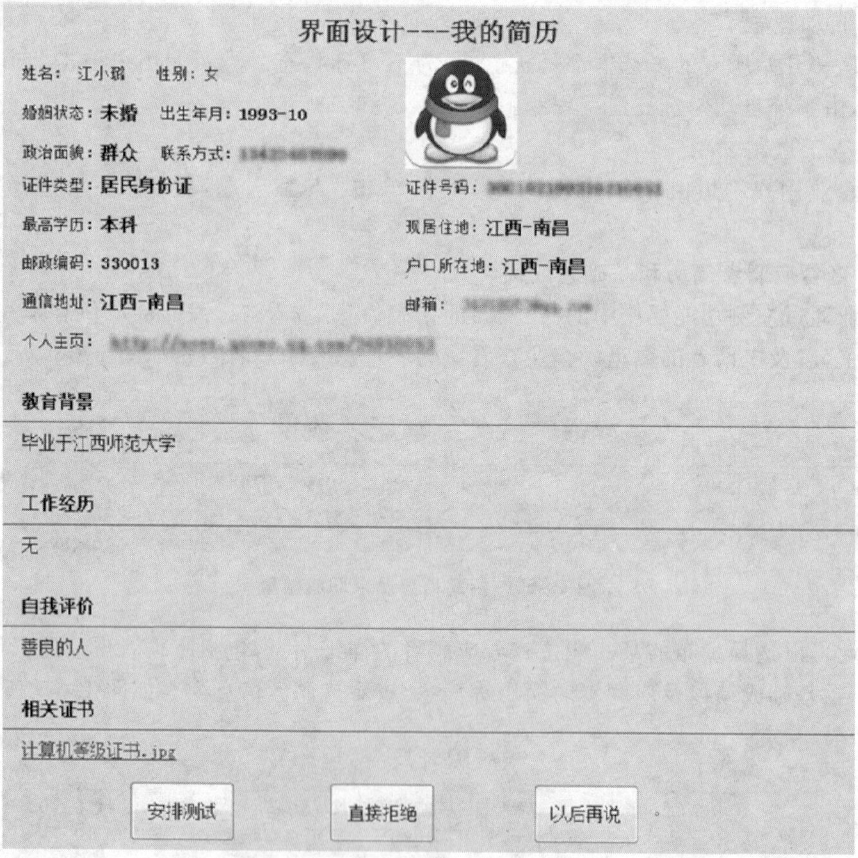

图 19-53 企业查看应聘者简历

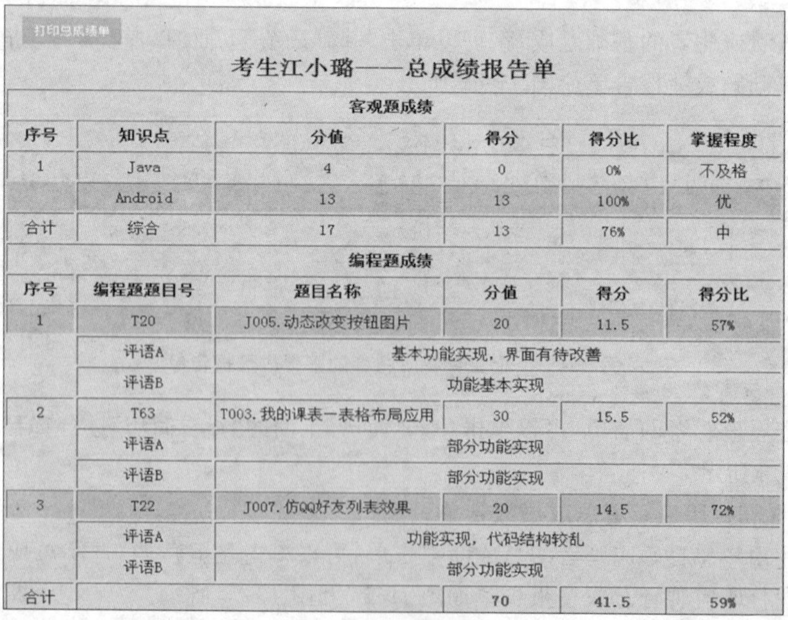

图 19-54 总成绩报告单

考生江小璐——总成绩报告单

客观题成绩					
序号	知识点	分值	得分	得分比	掌握程度
1	Java	4	0	0%	不及格
	Android	13	13	100%	优
合计	综合	17	13	76%	中

编程题成绩					
序号	编程题题目号	题目名称	分值	得分	得分比
1	T20	J005.动态改变按钮图片	20	11.5	57%
	评语A	基本功能实现，界面有待改善			
	评语B	功能基本实现			
2	T63	T003.我的课表—表格布局应用	30	15.5	52%
	评语A	部分功能实现			
	评语B	部分功能实现			
3	T22	J007.仿QQ好友列表效果	20	14.5	72%
	评语A	功能实现，代码结构较乱			
	评语B	部分功能实现			
合计			70	41.5	59%

图 19-55　打印总成绩单

19.4.2　应聘者操作流程

在使用 Android 程序员猎头系统之前，必须先输入相当数量的企业招聘信息和试题题库，以供应聘者浏览企业发布的招聘信息并测试自己对基础知识的掌握程度。求职者可以在不申请工作的情况下事先把简历填好，也可以先寻找工作，见到合适的职位申请，再选择简历一栏填写新的简历触发本功能。应聘者简历需要输入基本信息、工作经历等信息。应聘者操作流程如图 19-56 所示。

1. 职位查询

应聘者登录到招聘网站后可以搜索职位，系统添加了筛选条件，应聘者可以按照输入的职位名称进行搜索；也可以根据选择的职位类别和工作所在的省份搜索职位。应聘者查找职位如图 19-57 所示。

2. 填写简历

应聘者在注册并登录界面后，可以填写个人简历。应聘者可以单击左侧功能列表中的"填写简历"，进入基本信息页面。填写个人简历界面如图 19-58 所示。

3. 查看和修改简历

应聘者填写完简历后，单击"保存"按钮，简历的信息将被保存到数据库中。应聘者可单击我的简历列表对已生成的简历进行查看、更新和删除等操作。修改个人简历如图 19-59 所示。

4. 基础题的测试

应聘者单击"客观题测试"后，进入客观题测试列表界面。当应聘者想进行基础题的测试时，可以单击"添加客观题测试试卷"，此时在列表中将新添加一条测试记录。这里应聘者可以多次添加基础题试卷进行测试。在申请职位时，可以选择最优的成绩投递给企业。添加基础题测试试卷如图 19-60 所示。

Android 编程经典案例解析

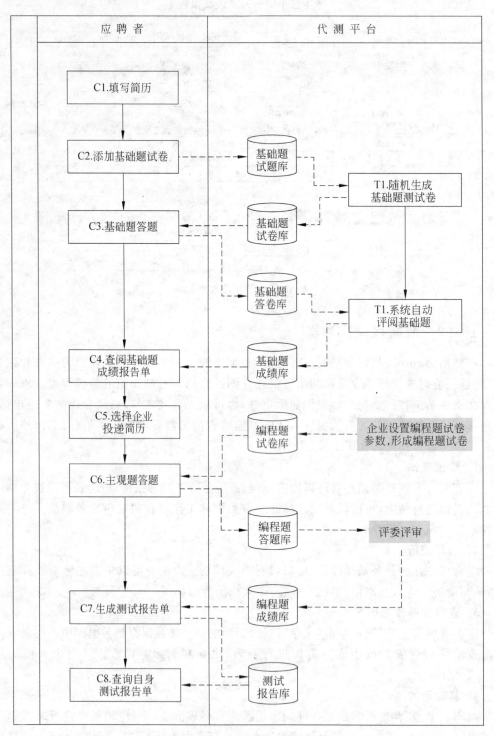

图 19-56 应聘者操作流程图

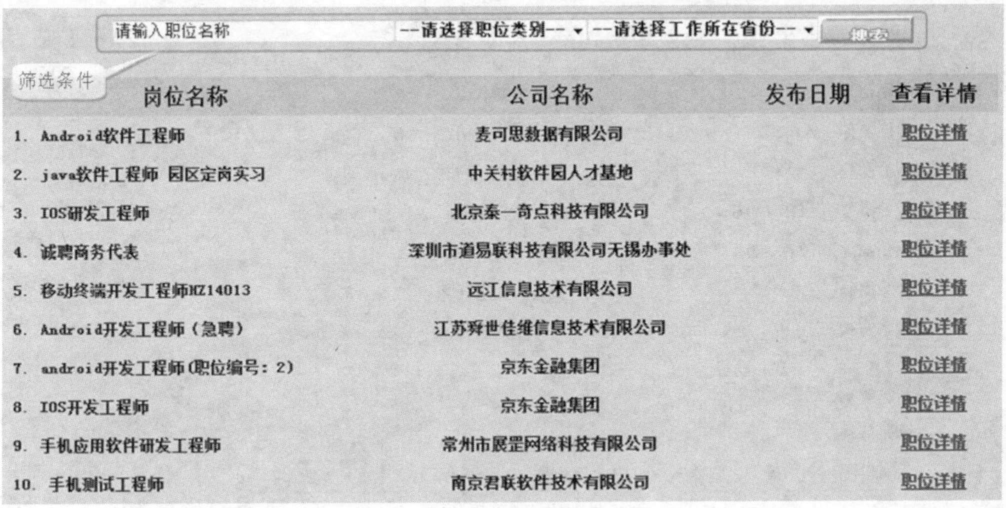
请输入职位名称
--请选择职位类别--
--请选择工作所在省份--
搜索
筛选条件
岗位名称
公司名称
发布日期
查看详情
1. Android软件工程师
麦可思数据有限公司
职位详情
2. java软件工程师 园区定岗实习
中关村软件园人才基地
职位详情
3. IOS研发工程师
北京秦一奇点科技有限公司
职位详情
4. 诚聘商务代表
深圳市道易联科技有限公司无锡办事处
职位详情
5. 移动终端开发工程师HZ14013
远江信息技术有限公司
职位详情
6. Android开发工程师（急聘）
江苏舜世佳维信息技术有限公司
职位详情
7. android开发工程师(职位编号：2)
京东金融集团
职位详情
8. IOS开发工程师
京东金融集团
职位详情
9. 手机应用软件研发工程师
常州市展罡网络科技有限公司
职位详情
10. 手机测试工程师
南京君联软件技术有限公司
职位详情

图 19-57　职位查询

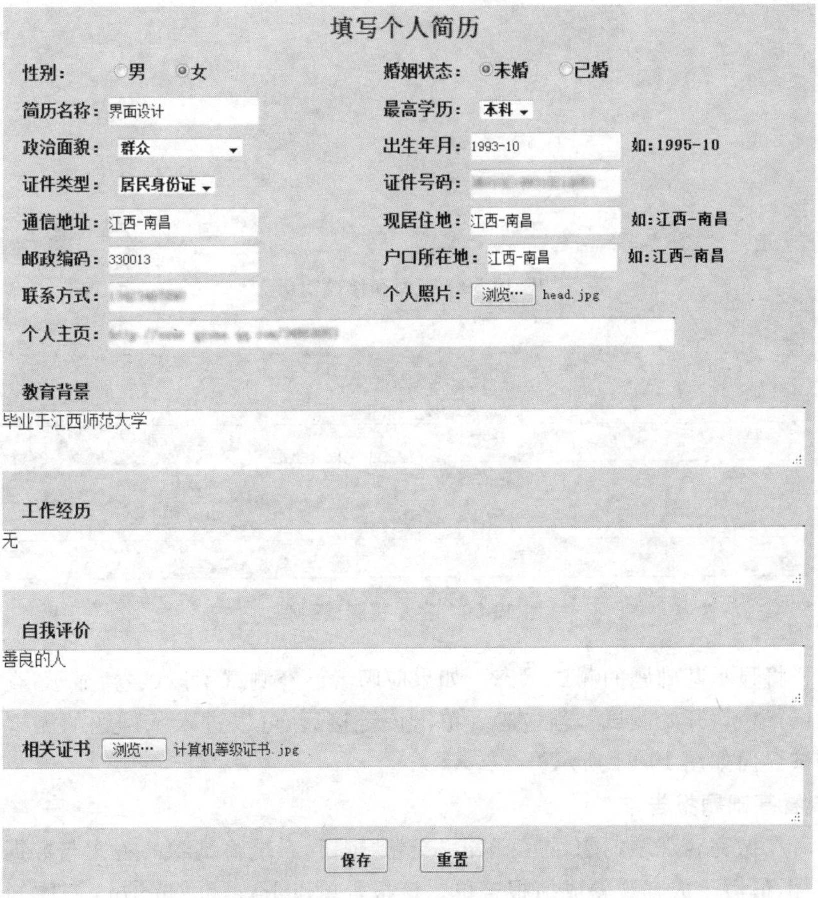
填写个人简历
性别：　男　女
婚姻状态：　未婚　已婚
简历名称：界面设计
最高学历：本科
政治面貌：群众
出生年月：1993-10　如:1995-10
证件类型：居民身份证
证件号码：
通信地址：江西-南昌
现居住地：江西-南昌　如:江西-南昌
邮政编码：330013
户口所在地：江西-南昌　如:江西-南昌
联系方式：
个人照片：浏览… head.jpg
个人主页：
教育背景
毕业于江西师范大学
工作经历
无
自我评价
善良的人
相关证书　浏览… 计算机等级证书.jpg
保存
重置

图 19-58　应聘者基本信息页面

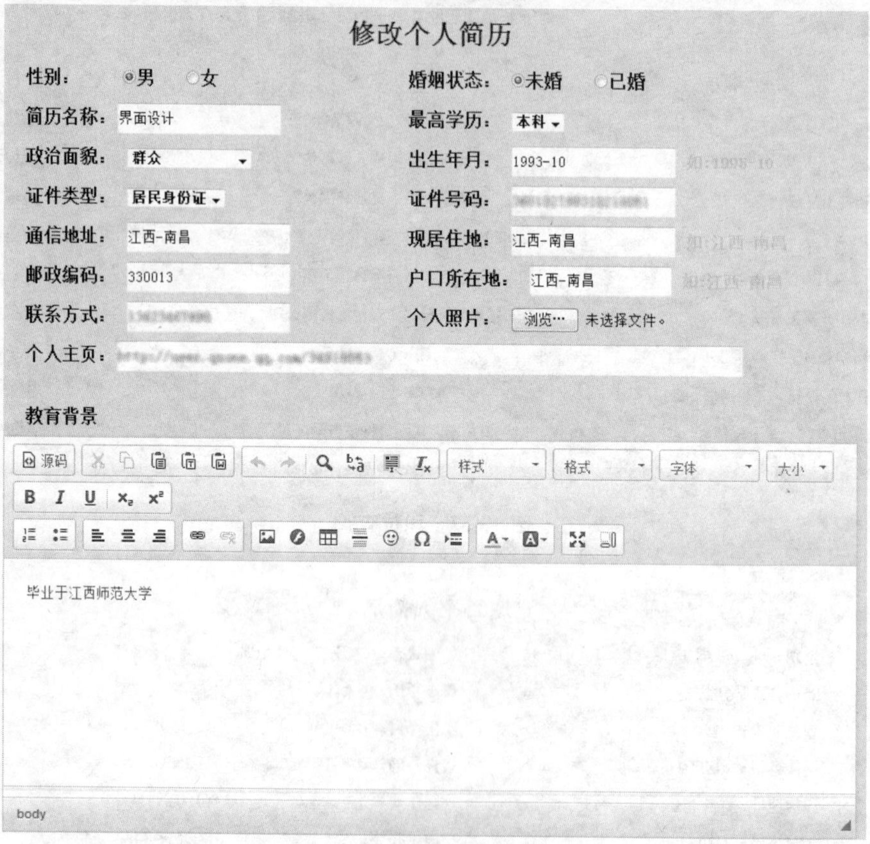

图 19-59　查看和修改简历

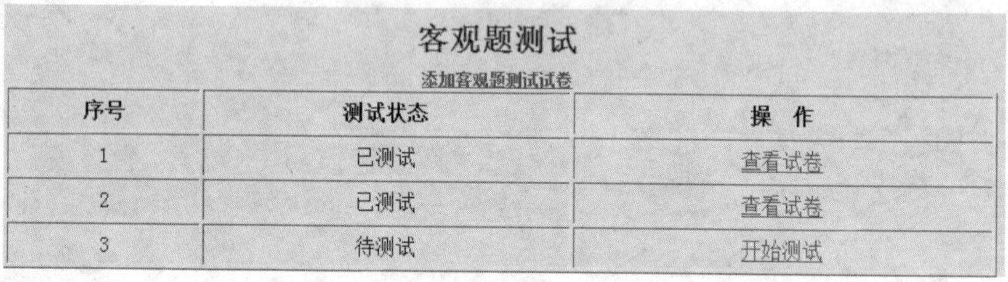

客观题测试

添加客观题测试试卷

序号	测试状态	操　作
1	已测试	查看试卷
2	已测试	查看试卷
3	待测试	开始测试

图 19-60　客观题测试列表

操作栏将显示基础题的测试状态。如果应聘者已经测试了该试卷，显示“查看试卷”；如果未测试，显示“开始测试”，当应聘者单击该链接后，可以进入基础题的测试界面。基础题的测试界面如图 19-61 所示。

5. 查看基础题报告单

应聘者在做完客观题的题目后，单击“交卷”按钮，系统将试卷的答案与数据库中的答案进行匹配，最后生成客观题成绩报告单。该报告单以 Java 和 Android 不同知识点的形式给出得分及得分比例，更能显示求职者对基础知识的掌握程度。基础题成绩报告单如图 19-62 所示。

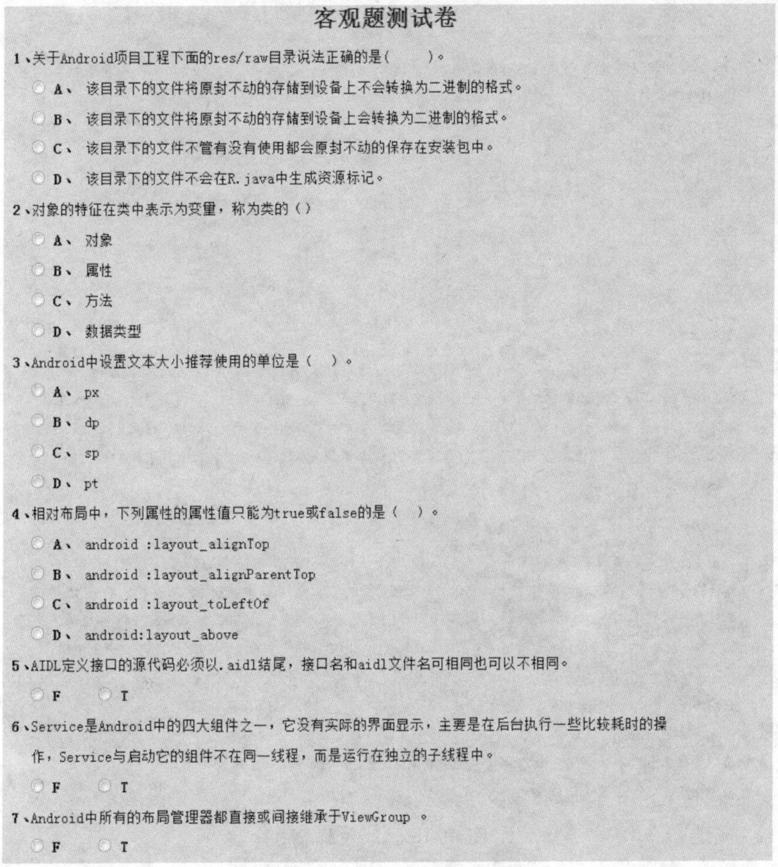

客观题测试卷

1、关于Android项目工程下面的res/raw目录说法正确的是(　　)。
　　A、该目录下的文件将原封不动的存储到设备上不会转换为二进制的格式。
　　B、该目录下的文件将原封不动的存储到设备上会转换为二进制的格式。
　　C、该目录下的文件不管有没有使用都会原封不动的保存在安装包中。
　　D、该目录下的文件不会在R.java中生成资源标记。

2、对象的特征在类中表示为变量，称为类的(　)
　　A、对象
　　B、属性
　　C、方法
　　D、数据类型

3、Android中设置文本大小推荐使用的单位是(　　)。
　　A、px
　　B、dp
　　C、sp
　　D、pt

4、相对布局中，下列属性的属性值只能为true或false的是(　　)。
　　A、android:layout_alignTop
　　B、android:layout_alignParentTop
　　C、android:layout_toLeftOf
　　D、android:layout_above

5、AIDL定义接口的源代码必须以.aidl结尾，接口名和aidl文件名可相同也可以不相同。
　　○F　　○T

6、Service是Android中的四大组件之一，它没有实际的界面显示，主要是在后台执行一些比较耗时的操作，Service与启动它的组件不在同一线程，而是运行在独立的子线程中。
　　○F　　○T

7、Android中所有的布局管理器都直接或间接继承于ViewGroup。
　　○F　　○T

图 19-61　客观题测试界面

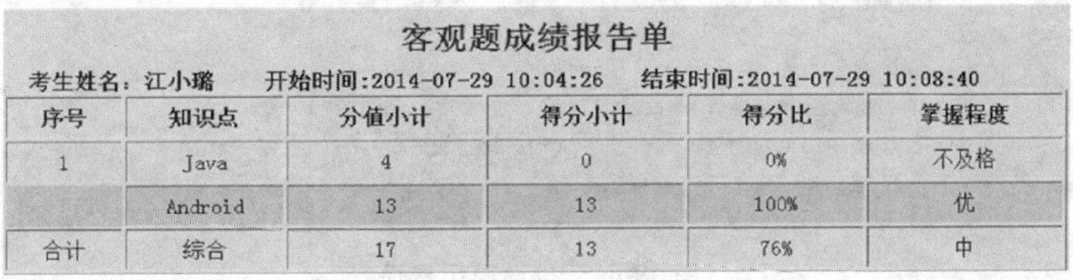

客观题成绩报告单

考生姓名：江小瑞　　开始时间：2014-07-29 10:04:26　　结束时间：2014-07-29 10:08:40

序号	知识点	分值小计	得分小计	得分比	掌握程度
1	Java	4	0	0%	不及格
	Android	13	13	100%	优
合计	综合	17	13	76%	中

图 19-62　客观题成绩报告单

6. 申请职位

应聘者在搜索职位时，如果找到适合自己的岗位，可以单击该职位所对应的"职位详情"查看职位的详细描述和应聘要求。职位详情如图 19-63 所示。

单击"立即申请"按钮后，系统将显示申请职位的名称、工作类型、工作地点及公司名称，并且提示应聘者选择简历和客观题成绩报告单。投递简历界面如图 19-64 所示。

7. 查看申请的职位信息

应聘者申请职位后，可以单击"我申请的工作"查看已申请职位的详细信息。已申请职位的信息列表如图 19-65 所示。

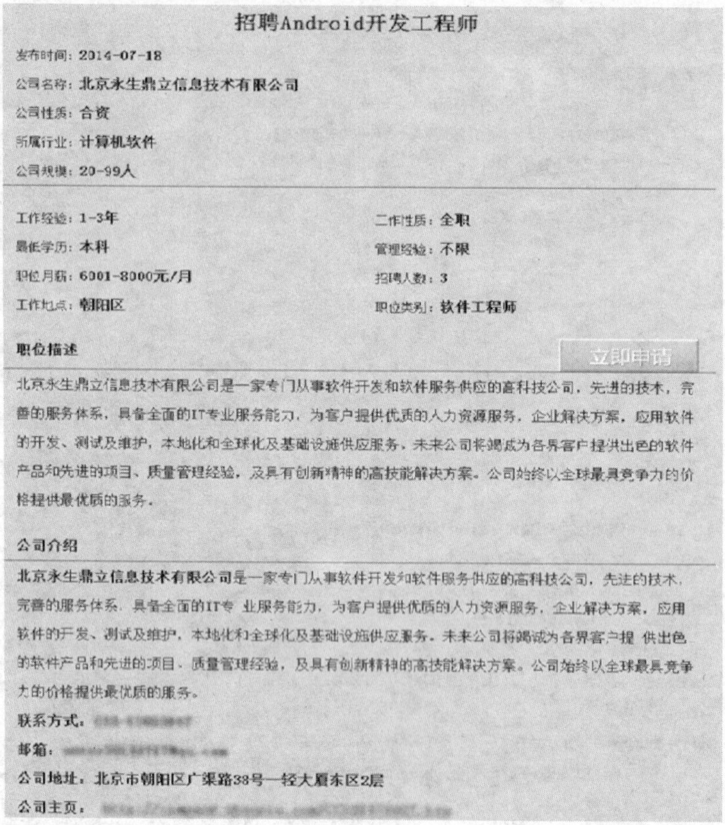

图 19-63　职位详情

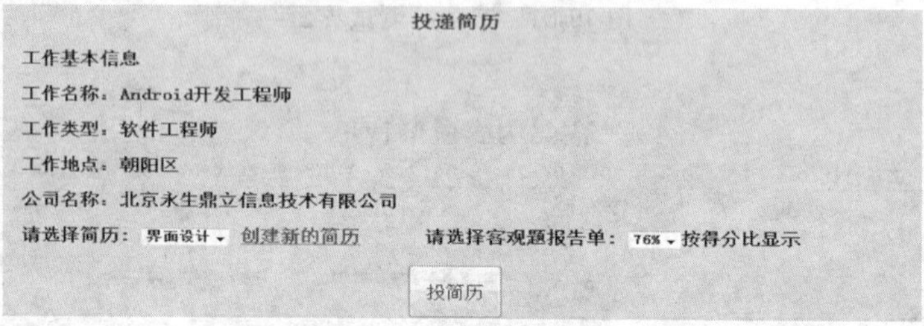

图 19-64　投递简历界面

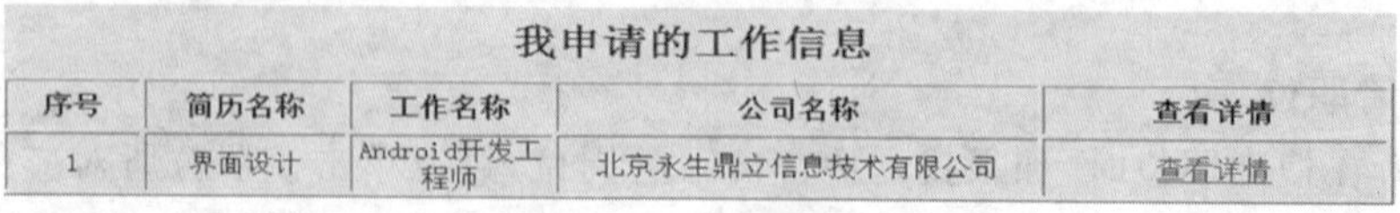

我申请的工作信息

序号	简历名称	工作名称	公司名称	查看详情
1	界面设计	Android开发工程师	北京永生鼎立信息技术有限公司	查看详情

图 19-65　申请的职位信息

8. 编程题的测试

应聘者申请职位后，如果企业用户已设置该职位的编程题试卷，并安排了应聘者进行测试，这时应聘者单击左侧功能模块中的“编程题测试”，将显示可测试的试卷信息。编程题测试信息列表如图 19-66 所示。

主观题测试

序号	工作名称	公司名称	测试状态	操作
1	Android开发工程师	北京永生鼎立信息技术有限公司	待测试	开始测试

图 19-66　编程题测试列表

当应聘者单击“开始测试”时，可以进入编程题测试列表界面进行编程题的测试。编程题的测试界面如图 19-67 所示。

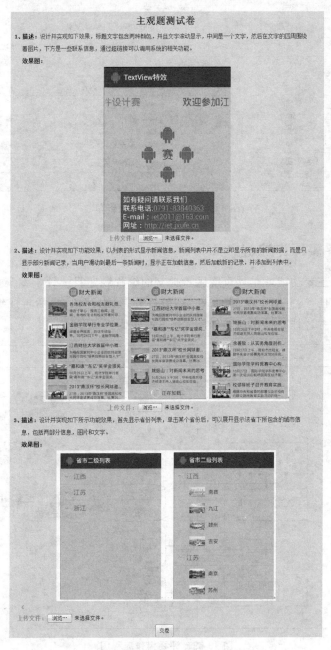

图 19-67　编程题测试界面

9. 查看编程题报告单

在应聘者提交编程题试卷后，等待评委评分。在评委评分完毕后，即可生成编程题报告单进行查看。编程题成绩报告单如图 19-68 所示。

编程题成绩报告单					
考生姓名：江小璐　　开始时间：2014-07-18 09:38:22　　结束时间：2014-07-29 10:29:15					
序号	编程题题目号	题目名称	分值	得分	得分比
1	T22	J007.仿QQ好友列表效果	20	14.5	72%
	评语A	功能实现，代码结构较乱			
	评语B	部分功能实现			
2	T20	J005.动态改变按钮图片	20	11.5	57%
	评语A	基本功能实现，界面有待改善			
	评语B	功能基本实现			
3	T63	T003.我的课表—表格布局应用	30	15.5	52%
	评语A	部分功能实现			
	评语B	部分功能实现			
合计			70	41	59%

图 19-68　编程题成绩报告单

19.4.3　评委操作流程

评审专家的主要功能就是评审应聘者提交的编程题题目。评审专家登录系统后可以查看自己的评审任务，并对分配到的试卷进行打分，给出评语，最后生成编程题报告单反馈给企业和应聘者。评委操作流程如图 19-69 所示。

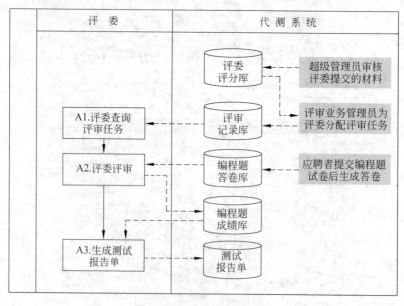

图 19-69　评委操作流程图

评委登录系统后，可以查看自己的评审任务。评委的评审任务列表如图 19-70 所示。

我的评审任务

导语：公平竞争为荣 携手目标达成

我的评审状态：　　　题目编号： 0

搜索

序号	题目编号	评审程序	姓名	评审分数	评　语	整体状态	我的状态
1	23	140195301123600150023.rar	assess01	10	功能实现	评审结束	已评审
2	70	140195301124200150070.rar	assess01	20	功能实现，界面良好	评审结束	已评审
3	71	140195301124700150071.rar	assess01	20	功能完全实现，界面有待优化	评审结束	已评审
4	20	140564960064000010020.zip	assess01	10	功能基本实现	评审结束	已评审
5	22	140564960068700010022.zip	assess01	15	实现程序功能，界面良好	评审结束	已评审
6	63	140564960068700010063.zip	assess01	15	部分功能实现，代码结构较乱	评审结束	已评审
7	20	140567327223100170020.zip	assess01	10	部分功能实现，代码结构有待优化	评审结束	已评审
8	22	140567327226200170022.zip	assess01	15	程序功能基本实现	评审结束	已评审
9	63	140567327226200170063.zip	assess01	15	部分功能实现，界面有待完善	评审结束	已评审
10	20	140591035016900020020.zip	assess01	12	部分功能实现，代码有待调整	评审结束	已评审
11	22	140591035018500020022.zip	assess01	10	部分功能实现	评审结束	已评审

图 19-70　评委的评审任务

1. 搜索评审任务

评委可以根据题目编号和评审状态筛选出需要评审的编程题。按“待评审”状态搜索到的评审任务如图 19-71 所示。

序号	题目编号	评审程序	姓名	评审分数	评　语	整体状态	我的状态
1	20	20140729text1.rar	assess01	0		未评审	待评审
2	22	20140729text2.rar	assess01	0		未评审	待评审
3	63	20140729text3.rar	assess01	0		未评审	待评审

图 19-71　搜索待评审记录

按题目编号搜索到的评审任务如图 19-72 所示。

2. 下载评审试卷和试卷参考答案

评审专家单击“待评审”进入评审界面，在评审前需要下载应聘者提交的编程题和参考答案，并保存在本地计算机上。下载链接如图 19-73 所示。

3. 评审

评审专家评审已经分配到的编程题。评审方案按实现的功能、界面和代码的编写情况进行打分，然后给出每道题的评语，以供应聘者和企业参考。评审界面如图 19-74 所示。

Android 编程经典案例解析

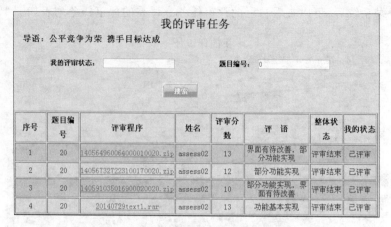

图 19-72　按题目编号搜索评审记录

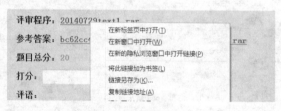

图 19-73　下载链接

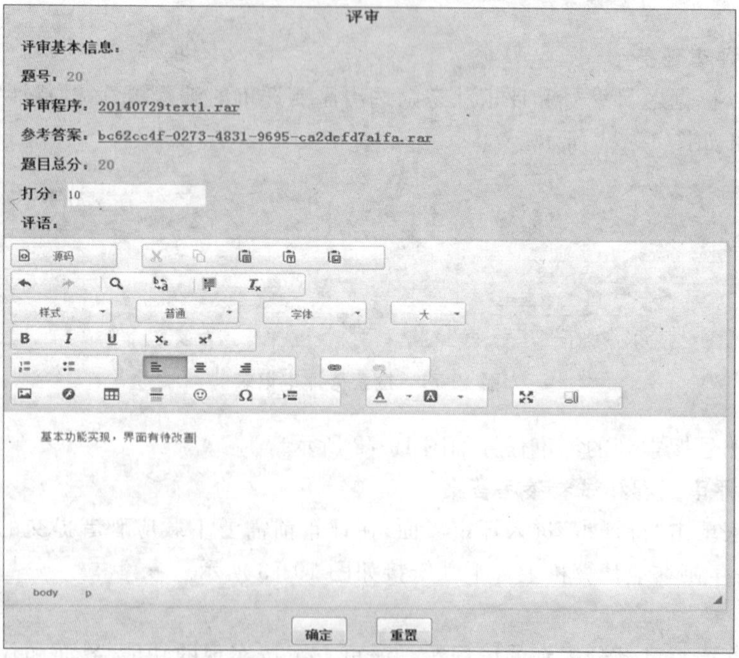

图 19-74　评审界面

19.4.4 超级管理员操作流程

超级管理员是 Android 程序员猎头系统的重要角色,功能包括审核企业用户和评审用户的资料,添加三类管理员,查看和更新用户的登录密码。该角色是该管理系统中必不可少的角色之一。超级管理员操作流程如图 19-75 所示。

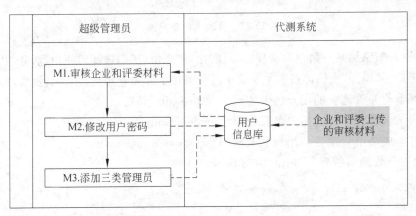

图 19-75　超级管理员操作流程图

1. 用户管理

1) 查看和更新用户登录密码

超级管理员在功能上可以管理所有用户,单击左侧功能模块的"用户管理"下的修改用户密码,界面中将显示所有用户的用户名和密码信息。超级管理员查看所有用户信息如图 19-76 所示。

	所有用户信息列表		
序　号	用户名	密　码	操　作
1	aaa	aaa	更新
2	bbb	bbb	更新
3	ccc	ccc	更新
4	ddd	ddd	更新
5	eee	eee	更新
6	assess01	assess01	更新
7	assess02	assess02	更新
8	assess03	assess03	更新
9	assess04	assess04	更新
10	assess05	assess05	更新

图 19-76　查看和更新用户登录密码

如果想修改某个用户的密码,可以单击"更新"按钮,然后输入新密码。修改密码如图 19-77 所示。

2) 添加三类管理员(评审业务、试卷管理、打印成绩单)

单击左侧功能模块中的"添加管理员操作",在添加管理员右侧显示添加管理员的信息

图 19-77　修改用户密码操作

输入表格。若想添加某一角色的管理员，只需在"选择角色"中选择对应角色，并填写相对应的数据，再单击"添加"按钮即可。本系统中已经添加了三类管理员，用户名及密码如下。

- 评审业务管理员的用户名及密码为 assessadmin01。
- 测试试卷管理员的用户名及密码为 paperadmin01。
- 打印成绩单管理员的用户名及密码为 reportadmin01。

添加三类管理员的界面如图 19-78 所示。

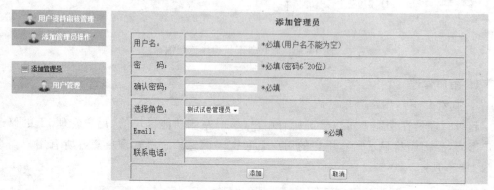

图 19-78　添加其他管理员

3）审核企业用户和评审用户资料

超级管理员进入登录界面后，即可看到审核的信息列表。其中有"审核资料"信息，可提供下载和查看。用户审核信息列表如图 19-79 所示。

超级管理员单击"查看注册信息"浏览用户的资料。如果评审企业用户的资料符合要求，管理员单击"通过审核"后，此时审核状态更改为"已审核"。用户审核界面如图 19-80 所示。

图 19-81 所示为企业用户注册时上传的审核材料（图示），管理员可以直接单击链接查看。

2. 试卷管理

试卷管理包括查看和删除试卷，查看、更新和删除题库中的试题等。良好的试卷管理，使得试题库中的试题抽取迅速。由于成绩报告单以知识点的形式统计分数，生成的成绩报告单也更加客观。试卷管理员操作流程如图 19-82 所示。

1）查看和删除试卷

应聘者添加基础题测试时，在试卷列表中会显示已经添加的试卷详细信息，包括试卷编号、试卷类型、生成时间。对于长期不使用的试卷，管理员可以单击"删除"按钮删除。查看和删除试卷列表的界面如图 19-83 所示。

2) 更新或删除题库

试卷管理员单击功能模块"题库管理"将按题型和知识点显示试题信息。试卷管理员可以对试题进行管理。测试题题库界面如图19-84所示。

序号	用户名	--请选择角色-- 筛选角色	审核资料	审核
1	assess01	评审	查看注册信息	已审核
2	assess02	评审	查看注册信息	已审核
3	assess03	评审	查看注册信息	已审核
4	assess04	评审	查看注册信息	已审核
5	assess05	评审	查看注册信息	已审核
6	enterprise01	企业用户	查看注册信息	已审核
7	enterprise02	企业用户	查看注册信息	已审核
8	enterprise03	企业用户	查看注册信息	已审核
9	enterprise04	企业用户	查看注册信息	已审核
10	enterprise	企业用户	查看注册信息	已审核
11	company	企业用户	查看注册信息	已审核
12	dahan	企业用户	查看注册信息	已审核
13	kkk	企业用户	查看注册信息	已审核
14	qqq	企业用户	查看注册信息	已审核
15	apple	企业用户	查看注册信息	已审核

需要审核的用户信息列表

图 19-79 审核企业用户和评审用户的资料

公司基本信息

公司名称：南昌倚动软件有限公司

公司性质：民营

公司规模：20-99人

联系方式：766018188

所属行业：计算机软件，教育培训

公司地址：江西省南昌市经济技术开发区江西财经大学蛟桥校区旁

公司主页：http://www.101ab.cn

公司介绍：
　　南昌倚动软件有限公司，是一家以移动互联网领域的软件开发、信息技术咨询服务为主体业务的高科技公司。该公司培养适应移动互联网行业要求，掌握3G手机软件设计、测试和管理基本技能，具备良好职业素养的熟练专业技术人才。

审核材料：
图15.jpg

通过审核

图 19-80 审核界面

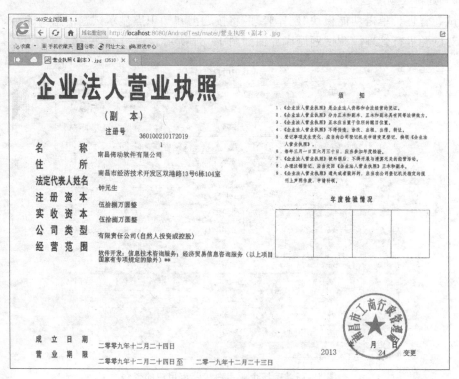

图 19-81　查看上传的审核材料

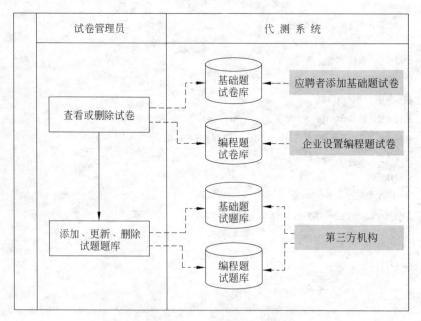

图 19-82　试卷管理员流程图

第 19 章 Android 程序员猎头系统

序号	试卷编号	试卷类型	生成时间	查看详情	删除
1	2c94cc03466ae61501466ae6e1890002	客观题试卷	2014-06-05 03:20:54	查看详情	删除
2	2c94cc03466ae61501466ae783ea0003	客观题试卷	2014-06-05 03:21:35	查看详情	删除
3	2c94cc03466ae61501466ae8927c0005	主观题试卷	2014-06-05 03:22:45	查看详情	删除
4	2c94cc03466f1a7901466f1afbbe0002	客观题试卷	2014-06-06 10:56:17	查看详情	删除
5	2c94cc03466f1a7901466f1b4a100003	客观题试卷	2014-06-06 10:56:37	查看详情	删除
6	2c94cc0347048148014704 81dcf60002	客观题试卷	2014-07-05 11:12:05	查看详情	删除
7	2c94cc0347048148014704 82cc440004	主观题试卷	2014-07-05 11:13:06	查看详情	删除
8	2c94cc0347052ef4014705333ce10001	主观题试卷	2014-07-05 02:25:49	查看详情	删除

图 19-83 试卷列表

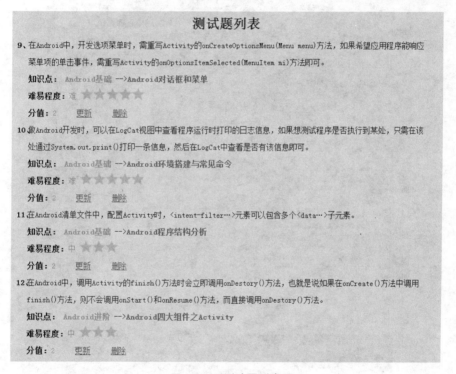

图 19-84 测试题列表

3. 评审管理

评审业务管理员由超级管理员添加角色并给予评审业务的权限。评审业务管理员具有分配编程题给某个评委的权力。评审业务管理员流程如图 19-85 所示。

评审业务管理员登录系统后,界面为评审记录列表,管理员可以查看已评审和未评审的相关记录。评审记录列表如图 19-86 所示。

在评审记录列表中,未被评审的编程题显示分配的状态为"分配评审员"。此时评审业务管理员可以单击该链接,进入分配界面(每题采取两位评委共同打分,得分为加权平均法),分配评委界面如图 19-87 所示。

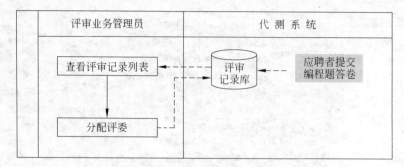

图 19-85　评审业务管理员流程图

序号	答卷编号	题目编号	评审程序	评审A	评审B	评审分数A	评审分数B	最后分数	评审状态	分配评审员
1	3	23	1401953011236001500023.rar	assess01	assess02	10	10	10.0	评审结束	已分配
2	3	70	1401953011242001500070.rar	assess01	assess02	20	30	25.0	评审结束	已分配
3	3	71	1401953011247001500071.rar	assess01	assess02	20	25	22.5	评审结束	已分配
4	6	23	1402023598162001600023.rar	assess03	assess04	16	10	13.0	评审结束	已分配
5	6	70	1402023598189001600070.rar	assess03	assess04	30	30	30.0	评审结束	已分配
6	6	71	14020235981930016001701.rar	assess03	assess04	10	15	12.5	评审结束	已分配
7	9	20	1405649600640000100020.zip	assess01	assess02	10	13	11.5	评审结束	已分配
8	9	22	1405649600687000100022.zip	assess01	assess02	15	16	15.5	评审结束	已分配
9	9	63	1405649600687000100063.zip	assess01	assess02	15	15	15.0	评审结束	已分配
10	11	20	1405673272231001700020.zip	assess01	assess02	10	12	11.0	评审结束	已分配
11	11	22	1405673272262001700022.zip	assess01	assess02	15	15	15.0	评审结束	已分配
12	11	63	1405673272262001700063.zip	assess01	assess02	15	16	14.0	评审结束	已分配

图 19-86　评审记录列表

图 19-87　分配评审员

4. 成绩单管理

成绩单管理员可以通过 Android 程序员猎头系统查看所有应聘者的考试成绩，成绩单管理员可以查看应聘者提交试卷的时间、应聘者的姓名和得分等详细信息。成绩单管理员流程如图 19-88 所示。

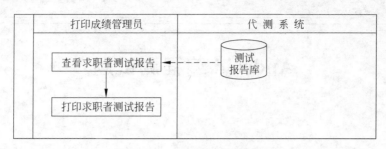

图 19-88　成绩单管理员流程图

单击"成绩管理",右侧会显示所有应聘者考试的成绩信息。成绩单列表如图 19-89 所示。

序号	答卷编号	提交时间	提交者	总分	得分	查看详情
1	12	2014-07-21 10:15:35	bbb	17	4	查看详情
2	13	2014-07-21 10:39:10	bbb	70	47	查看详情
3	14	2014-07-24 08:48:17	ddd	20	5	查看详情
4	19	2014-07-24 09:14:30	eee	16	3	查看详情
5	22	2014-07-29 09:59:54	程凯	19	14	查看详情
6	23	2014-07-29 10:08:40	江小璐	17	13	查看详情
7	24	2014-07-29 10:29:15	江小璐	70	41	查看详情

图 19-89　成绩单列表

单击"查看详情",可以查看应聘者试卷中每题的得分情况。编程题成绩报告单如图 19-90 所示。

编程题成绩报告单

考生姓名:江小璐　开始时间:2014-07-18 09:38:22　结束时间:2014-07-29 10:29:15

序号	编程题题目号	题目名称	分值	得分	得分比
1	T22	J007.仿QQ好友列表效果	20	14.5	72%
	评语A	功能实现,代码结构较乱			
	评语B	部分功能实现			
2	T20	J005.动态改变按钮图片	20	11.5	57%
	评语A	基本功能实现,界面有待改善			
	评语B	功能基本实现			
3	T63	T003.我的课表—表格布局应用	30	15.5	52%
	评语A	部分功能实现			
	评语B	部分功能实现			
合计			70	41	59%

图 19-90　编程题成绩报告单

Android 编程测试题

一、Android 环境搭建与程序结构分析

(1) Android 中启动模拟机(Android Virtual Device)的命令是(　　)。
　　A) adb　　　　B) android　　　　C) avd　　　　D) emulator
(2) Android 中完成模拟器文件与电脑文件间的复制以及安装应用程序的命令是(　　)。
　　A) adb　　　　B) android　　　　C) avd　　　　D) emulator
(3) Android 中创建模拟器的命令是(　　)。
　　A) android create avd －n（模拟器的名称）－t（android 版本）
　　B) adb create avd －n（模拟器的名称）－t（android 版本）
　　C) avd create avd －n（模拟器的名称）－t（android 版本）
　　D) emulator create avd －n（模拟器的名称）－t（android 版本）
(4) 下面关于 Android 项目工程下的 assets 目录和 res 目录的描述不正确的是(　　)。
　　A) assets 目录下可任意建立子文件夹,存放的资源都会原封不动地保存在安装包中,不会被编译成二进制
　　B) res 目录下的资源在打包时会判断是否被使用,未使用的资源将不会打包到 APK 中
　　C) assets 目录和 res 目录下的资源都会在 R.java 中生成资源标记
　　D) res 目录下只包括一些固定的子文件夹,不能任意创建子文件夹
(5) 关于 Android 项目工程下面的 res/raw 目录说法正确的是(　　)。
　　A) 该目录下的文件将原封不动地存储到设备上不会转换为二进制的格式
　　B) 该目录下的文件将原封不动地存储到设备上会转换为二进制的格式
　　C) 该目录下的文件不管有没有使用都会原封不动地保存在安装包中
　　D) 该目录下的文件不会在 R.java 中生成资源标记
(6) AndroidManifest 的文件扩展名是(　　)。
　　A) .jar　　　　B) .xml　　　　C) .apk　　　　D) .java
(7) 下列关于 Android 工程项目中的 AndroidManifest 清单文件说法正确的是(　　)。
　　A) AndroidManifest 清单文件是每个 Android 项目所必需的,它是整个 Android 应用的全局描述文件
　　B) AndroidManifest 文件说明了该应用的名称、所使用的图标以及包含的组件等

C) AndroidManifest 清单文件中包含了应用程序使用系统所需的权限声明,也包含了其他程序访问该程序所需的权限声明

D) AndroidManifest 清单文件的根元素是＜application＞,所包含的组件(如 Activity、Service 等)都包含在＜application＞元素内

(8) 下列关于 AndroidManifest 清单文件的内容描述正确的是( )。

A) 声明应用程序本身所需要的权限应放在＜application＞元素之内

B) 声明调用该应用程序所需的权限应放在＜application＞元素之外

C) 通过功能键,可以查看手机上的应用软件,功能清单中应用的标签可通过＜application＞元素的 android:label 属性进行设置

D) 通过功能键,可以查看手机上的应用软件,手机功能清单中应用的标签可通过主 Activity 的 android:label 属性进行设置

(9) Android 中的四大组件通常都会在 AndroidManifest 清单文件中进行注册,以下( )组件可以不在清单文件中注册也可以使用。

A) Activity B) Service

D) ContentProvider D) BroadcastReceiver

(10) 在下列描述中,不正确的是( )。

A) Android 应用的 gen 目录下的 R.java 被删除后会自动生成

B) Android 项目中的 res 目录是一个特殊目录,用于存放应用中的各种资源,命名规则可以支持大小写字母(a~z,A~Z)、数字(0~9)以及横线(_)

C) AndroidManifest 清单文件是每个 Android 项目必须有的,是项目应用的全局描述,通过包名+组件名可以指定组件的完整路径

D) Android 项目的 assets 和 res 目录都能存放资源文件,但是与 res 不同的是 assets 支持任意深度的子目录,在它里面的文件不会在 R.java 中生成任何资源 ID

(11) 下面( )不属于 Android 体系结构中的应用程序层。

A) 电话簿 B) 日历 C) SQLite D) SMS 程序

(12) 在清单文件中注册组件时,以下配置不正确的是( )。

A) `<activity android:name=".MyActivity">`
    `<intent-filter>`
        `<action android:name="iet.jxufe.cn.action.View"/>`
    `</intent-filter>`
`</activity>`

B) `<service android:name=".MyService"></service>`

C) `<provider android:name=".MyProvider"></provider>`

D) `<receiver android:name=".MyReceiver">`
    `<intent-filter>`
        `<action android:name="iet.jxufe.cn.receiver.myReceiver"/>`
    `</intent-filter>`

        </receiver>

二、Android 界面编程

(1) 在以下控件中,不是直接或间接继承自 ViewGroup 类的是(　　)。
   A) GridView　　　B) ListView　　　　C) ImageView　　　D) ImageSwitcher

(2) 以下不属于 Android 中的布局管理器的是(　　)。
   A) FrameLayout　　B) GridLayout　　　C) BorderLayout　　D) TableLayout

(3) 在 Android 中设置文本大小推荐使用的单位是(　　)。
   A) px　　　　　　B) dp　　　　　　　C) sp　　　　　　　D) pt

(4) 下列关于 TextView 和 ImageView 的说法正确的是(　　)。
   A) TextView 主要用于显示文字,可对文字大小、颜色等进行设置,TextView 除了设置背景图片外,不能在其上显示图片
   B) ImageView 主要用于显示图片,可设置图片的来源、缩放类型等,ImageView 上不能显示文字
   C) ImageView 从 TextView 继承而来,是对 TextView 的扩展
   D) 在 ImageView 标签中设置 android:text 属性时会直接报错

(5) 在下列选项中,前后两个类不存在继承关系的是(　　)。
   A) TextView、AutoCompleteTextView　　B) TextView、Button
   C) ImageView、ImageSwitcher　　　　　D) ImageView、ImageButton

(6) 在水平线性布局中,通过设置以下(　　)属性可以使得控件的宽度成一定的比例。
   A) android:layout_width　　　　　　　B) android:layout_weight
   C) android:layout_margin　　　　　　 D) android:layout_gravity

(7) 在下列属性中,不属于 EditText 文本编辑框的属性的是(　　)。
   A) android:inputType　　　　　　　　B) android:hint
   C) android:scaleType　　　　　　　　D) android:minLines

(8) 下列关于线性布局的描述正确的是(　　)。
   A) 水平线性布局中所有的控件都是按照水平方向一个挨着一个排列的,超出屏幕的宽度后,将会自动生成水平滚动条,拖动滚动条可查看其他控件
   B) 水平线性布局中所有的控件都是按照水平方向一个挨着一个排列的,超出屏幕的宽度后,将会自动换行显示其他控件
   C) 水平线性布局中所有的控件都是按照水平方向一个挨着一个排列的,超出屏幕的宽度后,将不会显示多余的控件
   D) 水平线性布局中所有的控件都是按照水平方向一个挨着一个排列的,超出屏幕的宽度后再添加控件,程序运行时将报错

(9) 下列关于表格布局的描述不正确的是(　　)。
   A) 表格布局从线性布局继承而来
   B) 表格布局中可明确指定包含多少行多少列
   C) 在表格布局中,可设置某一控件占多列

D) 如果直接向表格布局中添加控件，而不是在 TableRow 中添加，则该控件将单独占一行

(10) 在表格布局中，设置某一列为可收缩列的正确方法是（　　）。
　　A) 设置 TableLayout 的属性：android:stretchColumns="x"，x 表示列的序号
　　B) 设置 TableLayout 的属性：android:shrinkColumns="x"，x 表示列的序号
　　C) 设置具体列的属性：android:stretchable="true"
　　D) 设置具体列的属性：android:shrinkable="true"

(11) 在相对布局中，如果想让一个控件居中显示，则可设置该控件的（　　）。
　　A) android:gravity="center"
　　B) android:layout_gravity="center"
　　C) android:layout_centerInParent="true"
　　D) android:scaleType="center"

(12) 在相对布局中，属性值只能为 true 或 false 的是（　　）。
　　A) android:layout_alignTop
　　B) android:layout_alignParentTop
　　C) android:layout_toLeftOf
　　D) android:layout_above

(13) 在以下方法中，可以成功地将 ImageButton 的背景设为透明的是（　　）。
　　A) 设置 ImageButton 的 android:alpha 的属性值为 0
　　B) 设置 ImageButton 的 android:alpha 的属性值为 255
　　C) 设置 ImageButton 的 android:background 的属性值为 #ffffffff
　　D) 设置 ImageButton 的 android:background 的属性值为 #00000000

(14) 假设某张图片的大小为 1200×1200，现需将其显示在一个 300×200 的 ImageView 上，如果设置该 ImageVIew 的 scaleType 属性的值为 fitCenter，则图片的缩放比例为（　　）。
　　A) 横轴缩放比例为 4，纵轴缩放比例为 6
　　B) 等比例缩放，缩放比例为 6
　　C) 横轴缩放比例为 6，纵轴缩放比例为 4
　　D) 等比例缩放，缩放比例为 4

(15) 为下拉列表自定义 Adapter，即写一个类继承自 BaseAdapter 时，必须重写父类中的一些方法，以下（　　）方法不是必需的。
　　A) getCount()　　　　　　　　B) getView()
　　C) getItem()　　　　　　　　 D) getDropDownView()

(16) Android 中包含了很多 Adapter 的相关类，在下列选项中，不是从 BaseAdapter 继承而来的类是（　　）。
　　A) ArrayAdapter　　　　　　　B) SimpleAdapter
　　C) CursorAdapter　　　　　　 D) PagerAdapter

(17) 以下关于 SimpleAdapter 构造方法中参数的描述不正确的是（　　）。

A) 第一个参数为 Context 上下文对象,通常只需要传入当前的 Activity 对象

B) 第二个参数为列表的数据来源,既可以是一个数组,也可以是一个集合

C) 第三个参数为列表中每一项的布局文件,该布局中可以包含多个控件

D) 第四个参数与第五个参数之间存在一一对应的关系,根据第四个参数获取的数据,将会在第五个参数所指定的控件中显示,并且第五个参数中的元素必须在第三个参数指定的布局文件中

(18) AutoCompleteTextView(自动完成输入)控件可根据用户输入的内容从指定的数据源中匹配出所有符合条件的数据,并以下拉列表的形式显示,从而让用户进行选择。通过以下(　　)属性,可以设置弹出列表所需要用户输入的最少字符数。

　　A) android:completionThreshold

　　B) android:completionHint

　　C) android:dropDownVerticalOffset

　　D) android:dropDownHorizontalOffset

(19) 以下代表拖动条的控件是(　　)。

　　A) RatingBar　　　　　　　　B) ProgressBar

　　C) SeekBar　　　　　　　　　D) ScrollBar

(20) RatingBar 星级评分条中不能通过属性直接设置的是(　　)。

　　A) 五角星的个数　　　　　　B) 当前分数

　　C) 分数的增量　　　　　　　D) 五角星的色彩

(21) ScrollView 垂直滚动条中,最多可直接包含(　　)个子控件。

　　A) 0　　　　B) 1　　　　C) 2　　　　D) 无数

(22) 在以下控件中,不是从 Button 继承而来的是(　　)。

　　A) ImageButton　　　　　　　B) RadioButton

　　C) CheckBox　　　　　　　　D) ToggleButton

三、Android 对话框与菜单

(1) 下列关于 AlertDialog 的描述不正确的是(　　)。

　　A) AlertDialog 的 show()方法可创建并显示对话框

　　B) AlertDialog.Builder 的 create()和 show()方法都返回 AlertDialog 对象

　　C) AlertDialog 不能直接用 new 关键字构建对象,而必须使用其内部类 Builder

　　D) AlertDialog.Builder 的 show()方法可创建并显示对话框

(2) 在构建 AlertDialog 时需要借助其内部类 Builder,Builder 类中包含了很多方法,在下列方法中,方法的返回类型与其他项不同的是(　　)。

　　A) create()　　B) setMessage()　　C) setView()　　D) setAdapter()

(3) AlertDialog 对话框中的按钮最多可以有(　　)个。

　　A) 1　　　　B) 2　　　　C) 3　　　　D) 无数

(4) 在自定义对话框时,将 View 对象添加到当前对话框中的方法是(　　)。

　　A) setDrawable()　　　　　　B) setContent()

　　C) setAdapter()　　　　　　　D) setView()

(5) 在Android中如果需要创建选项菜单,则必须重写Activity的(　　)方法。

　　A) onCreateOptionsMenu()　　　　B) onCreateContextMenu()

　　C) onOptionsCreateMenu()　　　　D) onContextCreateMenu()

(6) 在菜单资源文件中,无法识别(　　)标签。

　　A) <menu>　　B) <item>　　C) <submenu>　　D) <group>

(7) 为某个菜单项创建子菜单的方法是(　　)。

　　A) add　　B) addMenu　　C) addSubMenu　　D) addMenuItem

(8) 以下事件处理方法,不适合处理选项菜单项的单击事件的是(　　)。

　　A) 使用onOptionsItemSelected(MenuItem item)方法处理

　　B) 使用onContextItemSelected(MenuItem item)方法处理

　　C) 使用OnMenuItemClickListener的onMenuItemClick(MenuItem item)方法处理

　　D) 使用onMenuItemSelected(int featureId,MenuItem item)方法处理

四、Android事件处理

(1) 在Android的事件处理机制中,基于监听的事件处理机制实现的基本思想应用了设计模式中的(　　)。

　　A) 观察者模式　　B) 代理模式　　C) 策略模式　　D) 装饰者模式

(2) 为复选框CheckBox添加监听是否选中的事件监听器,使用的方法是(　　)。

　　A) setOnClickListener

　　B) setOnCheckedChangeListener

　　C) setOnMenuItemSelectedListener

　　D) setOnCheckedListener

(3) 使用异步任务处理耗时操作时,Android系统为我们提供了AsyncTask抽象类,在继承该类时必须实现AsyncTask中的(　　)方法。

　　A) onPreExecute()　　　　B) doInBackground()

　　C) onPostExecute()　　　　D) onProgressUpdate()

(4) 在使用异步任务处理耗时操作时,以下方法中不能更改界面组件显示的是(　　)。

　　A) onPreExecute()　　　　B) doInBackground()

　　C) onPostExecute()　　　　D) onProgressUpdate()

(5) 在以下创建Message对象的语句中,不正确的是(　　)。

　　A) Message msg=new Message();

　　B) Message msg=Message.obtain();

　　C) Message msg=Message.obtain(Message message);

　　D) Message msg=Message.copyFrom(Message message);

五、Android中的资源定义

(1) 以下文件放入Android项目的res/drawable文件夹下,会直接报错或者不能在R.java文件中生成成员变量的是(　　)。

　　A) aaa.xml　　B) bbb.JPG　　C) CCC.jpg　　D) ddd.eee.jpg

(2) 以下文件放入 Android 项目的 res/drawable 文件夹下,不会报错的是(　　)。
　　A) my_picture.PNG　　　　　　　　B) myDog.JPG
　　C) myCat.png　　　　　　　　　　 D) 9_dog.jpg
(3) 以下选项中,不能表示合法颜色值的是(　　)。
　　A) #ggg　　　B) #ffff　　　C) #eeeeee　　　D) #dddddddd
(4) 使用 Android 中 Canvas 类的 drawRect(10,10,20,20,new Paint())绘制矩形,此矩形的面积是(　　)。
　　A) 100　　　B) 200　　　C) 300　　　D) 400
(5) Android 应用中定义了一些资源常量,通常放在<resources>标签下,下列(　　)不属于<resource>标签的子标签。
　　A) <string>　　　　　　　　　　　B) <color>
　　C) <drawable>　　　　　　　　　　D) <object-array>
(6) 下面自定义 style 的方式正确的是(　　)。
　　A) <resources>
　　　　　　<style name="myStyle">
　　　　　　　　<item name="android:layout_width">match_parent</item>
　　　　　　</style>
　　　　</resources>
　　B) <style name="myStyle">
　　　　　　<item name="android:layout_width">match_parent </item>
　　　　</style>
　　C) <resources>
　　　　　　<item name="android:layout_width">match_parent </item>
　　　　</resources>
　　D) <resources>
　　　　　　<style name="android:layout_width">match_parent </style>
　　　　</resources>
(7) 在 Android 中,ImageButton 按钮的图片不仅可以是 jpg、png 格式的图片文件,也可以是 XML 文件定义的图片,如果需要定义一个随着按钮状态变化的 XML 文件图片,则该文件的根元素是(　　)。
　　A) <animation-list>　　　　　　　B) <layer-list>
　　C) <selector>　　　　　　　　　　D) <shape>
(8) 下列(　　)类 Drawable 对象,可以实现图片徐徐展开的效果。
　　A) StateListDrawable　　　　　　 B) LayerDrawable
　　C) ShapeDrawable　　　　　　　　 D) ClipDrawable
(9) 在 Android 中既可以在程序中定义动画,也可以在 XML 文件中定义动画,在 XML 文件中定义逐帧动画的根元素是(　　)。
　　A) <set>　　　　　　　　　　　　 B) <animation-list>

C) <layer-list> D) <selector>

(10) 下面是一个 XML 资源定义文件,关于这个文件的描述正确的是(　　)。

```
<?xml version="1.0" encoding="utf-8"?>
<shape xmlns:android="http://schemas.android.com/apk/res/android"
    android:shape="line">
    <stroke
        android:color="@color/gray"
        android:dashWidth="5dp"
        android:dashGap="3dp"/>
</shape>
```

A) 这个 shape 文件是画一个宽为 5dp、高为 3dp 的色块

B) 这个 shape 文件是画一个宽从 5dp 到 3dp 的等腰梯形

C) 这个 shape 文件是画一个底为 5dp、高为 3dp 的等腰三角形

D) 这个 shape 文件是画一条虚线,实线段 5dp,间隔 3dp

六、Android 四大组件之 Activity

(1) 下列不属于 Activity 的 launchMode 属性的值的是(　　)。

A) singleStack　　B) singleTop　　C) singleTask　　D) singleInstance

(2) 在下列选项中,(　　)不是 Activity 启动的方法。

A) startActivity　　　　　　　　B) goToActivity

C) startActivityForResult　　　　D) startActivityFromFragment

(3) 假设设置 MainActivity 的 lauchMode 属性值为 singleInstance,并且 MainActivity 已经存在于栈中,此时当前的 Activity 跳转到 MainActivity,将会首先调用 MainActivity 的(　　)方法。

A) onCreate()　　　　　　　　B) onResume()

C) onNewIntent()　　　　　　　D) onSaveInstanceState()

(4) 在下列方法中,(　　)不是 Activity 生命周期里的方法。

A) onCreate()　　B) onStart()　　C) onStop()　　D) onFinish()

(5) 在配置 Activity 时,下列(　　)是必不可少的。

A) android:name　　　　　　　B) <action.../>

C) <intent-filter.../>　　　　　　D) <category.../>

(6) 下列关于应用程序的入口 Activity 的描述,不正确的是(　　)。

A) 每个应用程序有且只有一个入口 Activity,没有入口 Activity 的应用,运行时将会报错

B) 入口 Activity 的<intent-filter.../>元素中可以有多个<action.../>标签

C) 入口 Activity 的<intent-filter.../>元素中可以有多个<category.../>标签

D) 入口 Activity 的<intent-filter.../>元素中必须有一个<action android:name="android.intent.action.MAIN"/>元素,并且有一个<category android:name="android.intent.category.LAUNCHER"/>元素

(7) 在清单文件中,配置 Activity 时,以下(　　)标签无法在<intent-filter.../>中识别。

　　A) <action...>　　　　　　　　　　B) <category.../>
　　C) <data.../>　　　　　　　　　　　D) <type...>

(8) 下列关于<intent-filter.../>标签的说法不正确的是(　　)。

　　A) 该标签内可以包含 0~N 个<action.../>子标签
　　B) 该标签内可以包含 0~N 个<category.../>子标签
　　C) 该标签内可以包含 0~N 个<data.../>子标签
　　D) 系统会根据该标签里的元素判断何时启动该组件

七、Android 中的数据存储

(1) 读取手机内置存储空间内文件的内容时首先调用的方法是(　　)。

　　A) openFileOutput()　　　　　　　B) read()
　　C) write()　　　　　　　　　　　　D) openFileInput()

(2) SharedPreferences 保存文件的路径和扩展名是(　　)。

　　A) /data/data/shared_prefs/*.txt
　　B) /data/data/package name/shared_prefs/*.xml
　　C) /mnt/sdcard/指定文件夹 指定扩展名
　　D) 任意路径/任意扩展名

(3) 对于一个已经存在的 SharedPreferences 对象 userPreference,想向其中存入一个字符串"name",userPreference 应该先调用(　　)方法。

　　A) edit()　　　B) save()　　　C) commit()　　　D) putString()

(4) 在以下数据类型中,不是 SQLite 内部支持的类型的是(　　)。

　　A) BLOB　　　B) INTEGER　　　C) VARCHAR　　　D) REAL

(5) 下列关于 SQLiteOpenHelper 的描述不正确的是(　　)。

　　A) SQLiteOpenHelper 是 Android 中提供的管理数据库的工具类,主要用于数据库的创建、打开、版本更新等,它是一个抽象类
　　B) 继承 SQLiteOpenHelper 的类,必须重写它的 onCreate()方法
　　C) 继承 SQLiteOpenHelper 的类,必须重写它的 onUpgrade()方法
　　D) 继承 SQLiteOpenHelper 的类,可以提供构造方法也可以不提供构造方法

(6) SQLiteOpenHelper 是 Android 中提供的管理数据库的工具类,用于管理数据库的创建、版本更新、打开等,它是一个抽象类。如果创建一个该类的子类,在以下方法中,(　　)不是必须要包含在新创建的类里的。

　　A) 构造方法　　　　　　　　　　　B) onCreate()
　　C) onUpgrade()　　　　　　　　　　D) getReadableDatabase()

(7) ContentProvider 是 Android 中的四大组件之一,写好 ContentProvider 后,需要在清单文件中进行配置,在配置<provider>标签时,以下(　　)属性是必需的。

　　A) android:name　　　　　　　　　B) android:authorities
　　C) android:exported　　　　　　　　D) A 和 B

(8) 阅读以下程序：

```
UriMatcher myUri=new UriMatcher(UriMatcher.NO_MATCH);
myUri.addURI("iet.jxufe.cn.providers.myprovider","person",1);
myUri.addURI("iet.jxufe.cn.providers.myprovider","person/#",2);
int result=myUri.match(Uri.parse("content://iet.jxufe.cn.providers
   .myprovider/person/10"));
```

程序执行结束后，result 的值为（　　）。

　　A）－1　　　　　　B）1　　　　　　C）2　　　　　　D）10

(9) ContentProvider 的作用是共享数据，暴露可供操作的接口，其他应用则通过（　　）来操作 ContentProvider 所暴露的数据。

　　A）ContentValues　　　　　　　　B）ContentResolver
　　C）URI　　　　　　　　　　　　　D）Context

(10) 如果一个应用通过 ContentProvider 共享数据，那么其他应用都可以对该数据进行操作，在读取数据时，可以使用 query 方法，该 query 方法是通过（　　）对象调用的。

　　A）ContentResolver　　　　　　　B）ContentProvider
　　C）SQLiteDatabase　　　　　　　 D）SQLiteHelper

(11) 在开发 Android 应用程序时，如果希望在本地存储一些结构化的数据，可以使用数据库，在 Android 系统中内嵌了一个小型的关系型数据库，即（　　）。

　　A）MySQL　　　　B）SQLite　　　　C）DB2　　　　D）Sybase

八、Android 四大组件之 Service 与 BroadcastReceiver

(1) 下列不属于 Service 生命周期的回调方法是（　　）。

　　A）onCreate()　　B）onBind()　　C）onStart()　　D）onStop()

(2) 通过以下（　　）方法可以提高 Service 的优先级。

　　A）setLevel()　　　　　　　　　　B）setPriority()
　　C）upgrade()　　　　　　　　　　D）startForeground()

(3) 关于 ServiceConnection 接口的 onServiceConnected()方法的触发条件描述正确的是（　　）。

　　A）bindService()方法执行成功后
　　B）bindService()方法执行成功同时 onBind()方法返回非空 IBinder 对象
　　C）Service 的 onCreate()方法和 onBind()方法执行成功后
　　D）Service 的 onCreate()和 onStartCommand()方法启动成功后

(4) 在开发 Service 组件时，需开发一个类使其继承系统提供的 Service 类，该类中必须实现 Service 中的（　　）方法。

　　A）onCreate()　　　　　　　　　　B）onBind()
　　C）onStartCommand()　　　　　　 D）onUnbind()

(5) 以下关于通过 startService()与 bindService()运行 Service 的说法不正确的是（　　）。

　　A）startService()运行的 Service 启动后与访问者没有关联，而 bindService()运

行的 Service 将与访问者共存亡

B) startService() 运行的 Service 将回调 onStartCommand() 方法,而 bindService() 运行的 Service 将回调 onBind() 方法

C) startService() 运行的 Service 无法与访问者进行通信、数据传递,bindService() 运行的 Service 可在访问者与 Service 之间进行通信、数据传递

D) bindService() 运行的 Service 必须实现 onBind() 方法,而 startService() 运行的 Service 没有这个要求

(6) 下列关于使用 AIDL 完成远程 Service 方法调用的说法不正确的是(　　)。

A) AIDL 定义接口的源代码必须以 .aidl 结尾,接口名和 aidl 文件名可以相同也可以不相同

B) aidl 的文件的内容类似 Java 代码

C) 创建一个 Service,在 Service 的 onBind() 方法中返回实现了 aidl 接口的对象

D) 在 AIDL 的接口和方法前不能加访问权限修饰符 public、private 等

(7) Android 手机启动后会发送一个广播,如果想让应用随开机启动,只需要在应用中接收该广播然后启动服务即可。该广播的 Action 的值是(　　)。

A) Intent.ACTION_BOOT_COMPLETED

B) Intent.ACTION_MAIN

C) Intent.ACTION_PACKAGE_FIRST_LAUNCH

D) Intent.ACTION_POWER_CONNECTED

(8) 关于广播接收器,以下描述正确的是(　　)。

A) 广播接收器只能在清单文件中注册

B) 广播接收器注册后不能注销

C) 广播接收器只能接收自定义的广播消息

D) 广播接收器可以在 Activity 中单独注册与注销

九、扩展

(1) 在 Android 中进行单元测试时,需要在清单文件中进行配置,以下说法错误的是(　　)。

A) 需要在 AndroidManifest 清单文件的<application>标签内配置 instrumentation

B) 需要在 AndroidManifest 清单文件的<manifest>标签内配置 instrumentation

C) 需要在 AndroidManifest 清单文件的<application>标签内配置 uses-library

D) 需要让测试类继承 AndroidTestCase 类

(2) 在 LogCat 视图中,Log 信息分为(　　)个级别。

A) 3　　　　　　B) 4　　　　　　C) 5　　　　　　D) 6

(3) 以下数据写法错误的是(　　)。

A) ["java","android"]

B) [{"java"},{"android"}]

C) {"id":1,"name":"java"}

D) [{"id":1,"name":"java"},{"id":2,"name":"android"}]

(4) MediaPlayer 在播放音/视频资源前,需要调用(　　)方法完成准备工作。
　　A) setDataSource()　B) prepare()　　C) begin()　　D) ready()
(5) 使用 MediaPlayer 播放存放在外部存储卡(sdcard)中的音乐文件的操作步骤是(　　)。
　　A) 使用 MediaPlayer.create()方法传入文件路径返回 MediaPlayer 对象,然后准备播放
　　B) 直接将音乐文件的路径传入 MediaPlayer 的构造方法中,然后准备播放
　　C) 先创建 MediaPlayer 对象,再调用 setDataSource 方法设置文件源,然后准备播放
　　D) A 和 C 都可以
(6) Android VM 虚拟机中运行的文件的扩展名为(　　)。
　　A).class　　　　B).apk　　　　C).dex　　　　D).xml

十、Android 编程实践题

(1) 请设计并实现如图 A-1 所示的界面效果。

界面要求:

① 界面整体采用垂直线性布局,将屏幕平分为上、下两个部分,上、下部分间距为 10dp,上部分的背景颜色为♯aabbcc,下部分的背景颜色为♯ccbbaa。

② 上半部分包含两个 TextView 控件,第一个 TextView 控件用于显示标题信息,在标题文字的左边和右边各有一个图标,图标为应用图标,标题文字与图片居中显示。标题信息与顶部的边距为 10dp,标题文字为"大学生手机软件设计赛",标题文字大小为 18sp。第二个 TextView 控件水平居中显示在上半部分的底端,文字内容为"丰厚大奖等你来拿\n 联系电话:15870219546\n 电子邮箱:86547632@qq.com \n 官方网站:www.10lab.cn\n 地点:中国-江西-南昌",文字颜色为白色(♯ffffff),文字大小为 16sp,背景颜色为蓝色(♯0000ff),边距为 5dp,文字内容中的电话、邮箱、网址以链接形式显示。

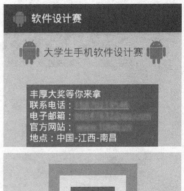

图 A-1　第 1 题的界面效果

③ 下半部分包含 4 个 TextView,大小分别为 160×160、120×120、80×80、40×40,单位为 dp,颜色分别为红色(♯ff0000)、绿色(♯00ff00)、蓝色(♯0000ff)、白色(♯ffffff),居中显示。

(2) 请设计并实现如图 A-2 所示的界面效果。

界面要求:

① 界面整体采用垂直线性布局,将屏幕平分为上、下两个部分,上、下部分间距为 10dp,上部分的背景颜色为♯aabbcc,下部分的背景颜色为♯ccbbaa。

② 上半部分包含两个 TextView 控件,第一个 TextView 控件用于显示标题信息,在标题文字的左边和右边各有一个图标,图标为应用图标,标题文字与图片居中显示。标题

信息与顶部的边距为10dp，标题文字为"大学生手机软件设计赛"，标题文字大小为18sp。第二个TextView控件水平居中显示在上半部分的底端，文字内容为"丰厚大奖等你来拿\n联系电话：15870219546\n电子邮箱：86547632@qq.com \n官方网站：www.10lab.cn\n地点：中国-江西-南昌"，文字颜色为白色(♯ffffff)，文字大小为16sp，背景颜色为蓝色(♯0000ff)，边距为5dp，文字内容中的电话、邮箱、网址以链接形式显示。

③ 下半部分包含5个ImageView，5个ImageView所显示的图片都为应用图标，以一个ImageView为中心，其他4个ImageView分别位于它的正上方、下方、左边、右边，整体处于下半部分的中间。

（3）请设计并实现如图A-3所示的界面效果。

图A-2 第2题的界面效果

图A-3 第3题的界面效果

界面要求：

① 界面中包含两个控件，即文本显示框 **TextView** 和列表 **ListView**。TextView用于显示标题信息，标题内容为"南昌景点介绍"，文字大小为24sp，背景颜色为♯ccbbaa，对齐方式为居中，边距为10dp。

② ListView显示所有景点信息，一项代表一个景点，每个景点包含三部分信息，即景点图片、景点名称、景点简介。ListView的背景颜色为♯aabbcc，项与项之间的分割线大小为2dp，颜色为灰色(♯aaaaaa)。

③ 在ListView的每一项中包含3个控件，一个ImageView用于显示景点图片，两个TextView分别显示景点名称、景点简介。其中，ImageView大小为100×75，景点名称文字大小为20sp，景点简介文字大小为12sp，颜色为♯0000ee，单行显示，当内容超过宽度时，省略后面的文字，以点代替。图片和相关文字介绍已放在BA3文件夹下。

（4）实现图片浏览功能，程序运行效果如图A-4所示。

界面要求：

界面中包含一个ImageView和3个Button，ImageView默认显示第一张图片，即

图 A-4　第 4 题程序的运行效果

file1.jpg。3 个按钮水平居中显示,按钮标签分别为"上一张"、"下一张"、"循环播放"。

功能要求:

① 为"上一张"、"下一张"按钮添加事件处理,使单击按钮后能够切换到上一张或下一张图片,事件处理方式不限,既可以使用绑定到标签也可以使用监听器。

② 为"循环播放"按钮添加事件处理,使单击该按钮后,程序能够自动地在多张图片间进行循环显示,每隔两秒进行切换,并且按钮文字发生变化,显示"停止播放",再次单击后,停止循环播放,按钮显示"循环播放"。相关图片资源已放在 BC1 文件夹下(关键:按钮在循环播放和停止播放之间切换、单击按钮能够循环播放和停止循环)。

(5) 实现拨号功能,程序运行效果如图 A-5 至图 A-8 所示。

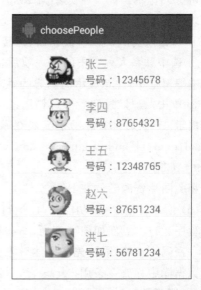

图 A-5　程序运行主界面　　　　图 A-6　显示联系人列表页面

图 A-7　显示选中的联系人　　　　　图 A-8　显示现在拨号页面

运行程序后,将显示图 A-5,单击图 A-5 中的"选择联系人"按钮后,会跳转到联系人列表页面,选中列表中的某一联系人后,会返回到主页面,并显示用户所选择的联系人,单击"拨号"按钮后,会调用系统的拨号功能,开始拨号。

界面要求:

首页中包含一个文本编辑框、两个按钮,其中,文本编辑框和"选择联系人"按钮水平摆放,文本编辑框的宽度为屏幕的宽度减去选择联系人按钮的宽度。

选择联系人页面中包含一个 ListView 列表控件,列表中的每一项包含三部分信息,即图标、姓名、电话号码。其中,图标大小为 50×50,姓名文字大小为 20sp,颜色为红色(♯ff0000),电话号码文字大小为 18sp,颜色为蓝色(♯0000ff)。

功能要求:

① 单击"选择联系人"按钮后,跳转到联系人列表页面。

② 选中联系人列表中的某一项后,会将联系人的姓名和号码返回给主页面,并显示在文本编辑框中,注意显示的内容为姓名:号码。

③ 单击"拨号"按钮后即可调用系统功能,进行拨号。

提示:调用系统拨号功能需提供相应的权限,拨打电话的权限为＜uses-permission android:name="android.permission.CALL_PHONE"/＞,系统中拨打电话的 Action 为 Intent.ACTION_CALL。

相关图片资源已放在 BC2 文件夹下。

(6) 实现注册、登录功能,程序运行效果如图 A-9 至图 A-12 所示。

程序运行主界面如图 A-9 所示,初始化时,输入任何账号和密码都不能登录,提示用户名和密码不正确,需要先注册然后才能登录。输入账号和密码后,单击"注册"按钮,即可向数据库中存放一条用户记录,然后输入该账号和密码,从数据库中查询,若发现存在该账号,即可登录;如果数据库中不存在该账号,则提示登录失败信息。

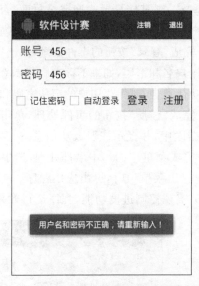

图 A-9　程序运行主界面　　　　图 A-10　用户名和密码不正确时弹出提示信息

图 A-11　用户注册成功时弹出提示信息　　　图 A-12　用户登录成功后切换页面显示信息

用户可以保存自己的登录信息,包括记住密码和自动登录,勾选"记住密码"后,下次登录时不需要输入账号和密码,勾选"自动登录"后,下次登录时直接跳转到欢迎界面。

界面要求:

此程序包含两个页面,一个是登录/注册主页面,另一个是登录成功显示用户信息的页面,这两个页面都已提供,在 BC3 文件夹下提供了一个可运行的不完整的应用程序,将其导入 Eclipse,在此基础上完善相关功能即可。

功能要求:

① 程序运行时,从 login.xml 文件中获取用户保存的相关信息,如果用户之前选择了

"自动登录",则直接显示欢迎登录页面,否则显示登录页面,然后判断是否记住密码,如果记住密码,则在相应的编辑框中显示上次输入的用户名和密码。

② 完成"登录"按钮的事件处理,单击"登录"按钮时,首先判断数据库中是否存在用户输入的用户名和密码,如果不存在则提示登录失败信息,如果存在则切换到欢迎界面,显示登录成功信息,同时保存用户的记住密码和自动登录的相关信息到 login.xml 文件中。

③ 完成"注册"按钮的事件处理,单击"注册"按钮时,将用户输入的用户名和密码保存到数据库中,并提示注册成功信息。

④ 完成菜单项的选中事件处理,选中注销菜单项时,取消自动登录,页面切换到主界面,选中退出菜单项时,退出应用程序。

(7) 实现控制进度功能,程序运行效果如图 A-13 所示。

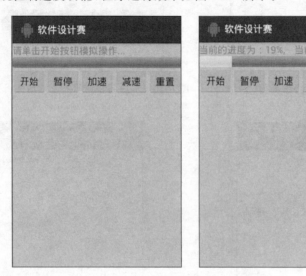

图 A-13 第 7 题程序的运行效果

界面要求:

界面设计已实现,在 BC3 文件夹下提供了一个可运行的不完整的应用程序,将其导入 Eclipse,在此基础上完善相关功能即可。直接运行该程序,将得到左图所示的效果。

功能要求:

① 完善"开始"按钮的事件处理方法,单击"开始"按钮后,进度条开始变化,在进度条上方的文字实时显示进度的信息,包括当前进度执行的百分比、进度递增的速度。

② 完善"暂停"按钮的事件处理方法,单击"暂停"按钮后,界面不再发生变化,保持为前一次的状态,此时单击除"暂停"之外的按钮可以执行相关操作,进度可以变化。

③ 完善"加速"按钮的事件处理方法,单击"加速"按钮后,在原有的速度之上加 5,显示的当前速度也随之发生变化。

④ 完善"减速"按钮的事件处理方法,单击"减速"按钮后,在原有的速度之上减 5,但最低速度为 1,显示的当前速度也随之发生变化。

⑤ 完善"重置"按钮的事件处理方法,单击"重置"按钮后,界面恢复到最开始的状态。
用户可在原有程序的基础上进行完善,也可以完全自主实现,达到功能即可。